JOHN ROBERTS' interest in geology dates from his ea~~~~ lived in Edinburgh under the shadow of Arthur's Seat, an ancient volca~~. After a good Scots education, he studied the subject at the University of Edinburgh, before undertaking research on the metamorphic rocks of the southwest Highlands of Scotland.

He then spent three years in the Middle East, teaching geology at universities in Saudi Arabia and the Lebanon, while travelling widely throughout the region, dodging the odd *coup d'état*. Returning to Britain in 1966, he devoted nearly every summer to studying the rocks of the Highlands for a series of scientific papers, while teaching structural geology and field-mapping at the University of Newcastle-upon-Tyne.

He took early retirement in 1986 to live near Tongue in the far north of Sutherland with his young family. As well as *The Highland Geology Trail*, originally published as a private venture by Strathtongue Press, he is author and photographer of *The Macmillan Field Guide to Geological Structures* (Macmillan, 1996), and *A Photographic Guide to Minerals, Rocks and Fossils* (New Holland, 1998). While living in Sutherland, he developed a strong interest in Highland history, culminating in the recent publication of *Lost Kingdoms: Celtic Scotland and the Middle Ages* (Edinburgh University Press, 1997). He now lives near St Andrews.

Cover Picture: Pinnacles formed by weathering and erosion of ancient lava-flows, where they have been affected by a huge landslip at the Old Man of Storr, north of Portree on the Isle of Skye.

Code of Conduct in the Field

Geologising in the field is not without its physical dangers, like any outdoor activity. The hazards can be reduced by wearing suitable clothing and footwear, preferably a pair of good boots. Wearing a safety helmet is a sensible precaution below loose cliffs. A compass and topographic map should always be carried, together with food for any emergency. Keep away from dangerous cliffs, and beware incoming tides. Always obey the Country Code: do not leave gates open; never climb over drystone walls; do not trample over crops; do not disturb farm animals; and do not leave any litter. Although trespass is not a criminal offence in Scotland, there is no right of access to the countryside. Permission should always be asked as appropriate from landowners, their agents and tenants. Respect any sporting interest in the land.

Never hammer rock exposures which show any particular feature of geological interest. This applies in particular to outcrops affected by weathering.

The use of geological hammers should be restricted in the interests of conservation. Much can be seen by using a hand lens on weathered surfaces, without the need for hammering. If thought absolutely necessary to examine fresh surfaces, only hammer loose specimens or fallen blocks. Keep collecting of rock, mineral and fossil specimens to an absolute minimum.

The
Highland Geology Trail

JOHN L. ROBERTS

Luath Press Limited

EDINBURGH

www.luath.co.uk

First published 1990 (Strathtongue Press)
Revised, updated and expanded 1998

The paper used in this book is recyclable. It is made from low chlorine
pulps produced in a low energy, low emission manner
from renewable forests.

Printed and bound by
Caledonian International Book Manufacturing Ltd., Glasgow

Typeset in 9 and 9.5 point Sabon by
S. Fairgrieve, Edinburgh, 0131 658 1763

Contents

Stratigraphic Column   vi

Foreword   vii

CHAPTER 1   Foundations of Geology   1

CHAPTER 2   Nature of the Geological Record   11

CHAPTER 3   Geological Record in the Scottish Highlands   21

CHAPTER 4   Inverness to John O'Groats   33

CHAPTER 5   John O'Groats to Cape Wrath   40

CHAPTER 6   Cape Wrath to Ullapool   49

CHAPTER 7   Clachtoll and Enard Bay   59

CHAPTER 8   Ullapool to Kyle of Lochalsh   63

CHAPTER 9   Isle of Skye   69

CHAPTER 10   Kyle of Lochalsh to Fort William   79

CHAPTER 11   Fort William to Oban   84

CHAPTER 12   Excursions from Oban   91

             Index of Geological Terms and Localities   102

Stratigraphic Column showing Geological Periods with Ages given in Millions of Years (my), and the Main Events in the Geological History of the Northwest Highlands

Quaternary	Pleistocene Glaciation (or Great Ice Age)
1.8 my	**Uplift and Erosion**
	Eruption of Basaltic Lavas from Central Volcanoes
Tertiary	followed by
	Intrusion of Central Igneous Complexes
70 my	
Cretaceous	Thin deposits of Chalk
140 my	
Jurassic	Deposition of Sandstones, Limestones and Shales
200 my	**Uplift and Erosion**
Triassic	
250 my	Deposition of New Red Sandstone (Permo-Triassic)
Permian	
300 my	**Uplift and Erosion**
Carboniferous	Gap in Geological Record
360 my	**Uplift and Erosion**
Devonian	Deposition of Old Red Sandstone
410 my	**Caledonian Earth-Movements on Moine Thrust**
Silurian	Gap in Geological Record
440 my	
Ordovician	Deposition of Cambrian Quartzite, Pipe Rock, Fucoid
520 my	Beds, Serpulite Grit and Durness Limestone of
Cambrian	Cambro-Ordovician Age
570 my	**Uplift and Erosion**
1000–800 my	Deposition of Stoer Group & Torridonian Sandstone
	Uplift and Erosion
1400 my	Metamorphism and Granite Intrusion (Laxfordian Gneisses)
2400 my	Intrusion of Scourie Dykes
2700 my	Deep-seated Metamorphism (Scourie Gneisses)

Foreword

ANYONE WANTING TO STUDY THE elements of geology can hardly do better than visit the Northwest Highlands of Scotland. A wide variety of different rock-types is found in this region, including nearly all the common types likely to be encountered in the field by the amateur geologist and natural historian. Instead of examining rocks as specimens in a museum, we can see how they occur in nature, and what serves to distinguish them from one another. But identifying rocks in the field is only a start. What soon becomes clear is the striking way in which geology underlies the scenic grandeur of the Northwest Highlands in all its diversity. No sooner do we pass from one geological domain to another than the scenery changes, often dramatically. The towering mountains of Torridonian Sandstone are quite different from the ice-scoured landscape of Lewisian Gneiss with its never-ending succession of rocky knolls and peaty lochans. Nowhere else in the British Isles is such an intimate relationship between geology and scenery to be seen. Understanding the geological foundations of the scenery must surely enrich our appreciation of the natural landscape, and nowhere more so than in the Highlands of Scotland.

Yet this is not all. Geology is essentially a historical science, allowing us to look into the 'abyss of time'. Rocks are not all formed at once, as the Old Testament might have us believe. They occur instead as the products of a long and complex history of geological events, spanning nearly three thousand million years in the case of the Northwest Highlands of Scotland. The realisation that geological time afforded almost 'no vestige of a beginning, no prospect of an end', was first appreciated by James Hutton (1726-97). He was a leading figure of the Scottish Enlightenment, whose *Theory of the Earth* provided the emerging science of geology with a solid foundation, based on field observations. He recognised in particular that geological history proceeds in a cyclical fashion, driven by deep-seated movements of the Earth's crust. There is thus a drama to geological events which can only be appreciated fully by looking at rocks in the field.

John L. Roberts
June 1998

Ordnance Survey Maps

No serious exploration of the countryside can be undertaken without the use of topographic maps. The Landranger Maps published on a scale of 1:50,000 by the Ordnance Survey are the best available. They allow the National Grid System to be used to maximum advantage. However, the Ordnance Survey also publishes Atlases on scales of 1:100,000 and 1:250,000, covering much more ground at much less expense, which provide a less satisfactory alternative.

Geological Survey Maps

The most useful geological map for the visitor is the Ten Miles to One Inch Geological Map of Great Britain (Solid Geology). Sheet One covers the Scottish Highlands. Other geological maps are available on scales of 1:250,000 and 1:50,000.

Geological Excursion Guides

Excursion Guide to the Geology of East Sutherland and Caithness. Geological Society of Aberdeen.
Lewisian and Torridonian Rocks of Northwest Scotland. Geologists' Association (Guide No. 21).
Excursion Guide to the Assynt District of Scotland. Edinburgh Geological Society.
Excursion Guide to the Isle of Skye. Geological Society of Glasgow.
Ardnamurchan: a Guide to Geological Excursions. Edinburgh Geological Society.
Guide to the Moine Geology of the Scottish Highlands. Scottish Academic Press.

British Regional Geology

The Northern Highlands of Scotland; The Grampian Highlands of Scotland; the Tertiary Volcanic Districts of Scotland. Published by the British Geological Survey.

Macmillan Field Guide to Geological Structures

The initials MFG and a number in the text refer to the localities where the corresponding photographs were taken for this field-guide by the author. It provides a complete description of all the geological structures likely to be encountered in the field. It may be complemented by John L. Roberts, *A Photographic Guide to Minerals, Rocks and Fossils.* New Holland, London, 1998.

CHAPTER I

Foundations of Geology

CHRISTIAN COSMOGONY ONCE PLACED humanity at the centre of the universe, inhabiting a solid earth created out of nothing by God. Around the earth revolved the sun, moon and the planets in their celestial spheres, while the fixed stars formed a more distant backdrop to the heavenly firmament. Only in the early 16th century did Copernicus reject such a cosmogony in favour of the heliocentric theory, first proposed by the Greek philosopher Aristarchus in the 3rd century BC. Not only did the earth and the planets revolve around the sun in their separate orbits, but later the sun was recognised as just another star, separated from the other stars by the vast distances of inter-stellar space.

Such a revolution in our understanding of the universe might well have revived Aristotle's idea that the earth was eternal, existing for all time. But deeply embedded in Christian cosmology was the paradigm of the Creation. Had not God created heaven and earth in six days, and on the sixth day made Adam and Eve in his own image? Even a detailed chronology could be computed from the Old Testament to show that the Creation occurred in 4004 BC, according at least to the calculations of Bishop Ussher of Armagh in 1664. Indeed, he concluded it had happened at 9 o'clock on the morning of the 26th of October in 4004 BC.

Yet such a chronology was difficult to reconcile with the natural world. Studying the great thicknesses of sedimentary strata exposed at the earth's surface, and the fossils preserved in such rocks as the remains of ancient life, made it ever more unlikely that the Earth's age could just be measured in thousands of years. Only if all living species, past and present, were created almost simultaneously according to the doctrine of divine creation, and only if the geological features of the earth and its landscape had arisen in rapid succession from a series of catastrophic events, such as the Biblical flood, lying quite outside the ordinary forces of nature, was it possible to believe literally in the testament of the Bible.

But as the scientific revolution reached its climax in the late 17th century, evidence slowly mounted that the earth's history must indeed be measured in millions of years. Even the very cooling of the earth from a molten state required such a long period of time, according to the calculations of Buffon (1707-80), based on experiments with cannonballs. Sedimentary rocks deposited upon the earth's surface could only have accummulated over great epochs of geological time, given that their deposition seemed to involve the very same processes as occurred at the present day. Likewise, the fossil sequences discovered in sedimentary rocks seemed to argue that life itself had a very ancient history. By the early 19th century, when geology first established itself as a scientific discipline in its own right, such a theory of gradual change, rather than the doctrine of catastrophism, became known as uniformitarianism. Its conception was forever linked with the name of James Hutton (1726-97).

Life of James Hutton

The son of a prosperous Edinburgh merchant, Hutton's intellectual background was broad and varied, embracing the pursuit of natural philosophy with practical achievement, especially in agriculture, as befitted a leading figure of the Scottish Enlightenment. He developed an early interest in chemistry, which eventually brought him considerable wealth, and then studied medicine at Edinburgh, Paris and Leyden. But he did not take up the practice of medicine, turning instead to agriculture, after he had inherited the arable farm of Sligh Houses near Duns in the Scottish Borders. There he put into effect the latest innovations of agricultural improvement, then at its height. Indeed, he even spent two years in Norfolk, studying the new methods of husbandry, while adding to his knowledge by travelling widely on the Continent, and elsewhere in England.

It was during these travels that James Hutton 'first began to study mineralogy, or geology....[becoming] very fond of studying the surface of the earth.... [and] looking with anxious curiosity into every pit, or ditch, or bed of a river that fell in his way'. Thereafter, he became an avid collector of geological specimens. By 1764, when he journeyed through the Highlands of Scotland as far north as Caithness with George Clark Maxwell, returning by way of the coast, Hutton's object was 'mineralogy, or rather geology.... which he was now studying with great attention'. Then, three years later, he returned to Edinburgh, where he spent the last thirty years of his life.

By then, Edinburgh was truly the 'Athens of the North', and all the major figures of the Scottish Enlightenment were among Hutton's friends, except perhaps for David Hume. He became a shareholder in the Forth and Clyde Canal, supervising its construction. After 1775 he devoted his life to scholarship and the company of his friends, while continuing to travel widely throughout the British Isles. He presented papers on meteorology, chemistry, natural history and geology to the Royal Society of Edinburgh, and took a prominent part in its affairs.

But it was only during the last five years of his life, when his health was failing, that he became a prolific author. After writing treatises upon physics, chemistry and philosophy, he finally published his *Theory of the Earth* in two volumes in 1795. It incorporated an earlier abstract of 1785, and a more extended account written in 1788, which had both been presented to the Royal Society of Edinburgh. A third volume was later discovered in manuscript form, and published in 1899.

Hutton's Theory of the Earth

Hutton's arguments were founded philosophically on the belief that there is a 'wise and beneficent design' to the natural world, since he declared at the outset that 'this globe of the earth is evidently made for man'. Indeed, his work is full of resounding assertions that nature in all its operations must be supervised by a benevolent Deity, intent upon creating and maintaining 'a world beautifully calculated for the growth of plants and nourishment of men and animals'. But equally he realised: 'a solid land could not have answered the purpose of a habitable world; for a soil is necessary to the growth of plants; and a soil is nothing but the material collected from the destruction of the solid land'.

Yet weathering and erosion in attacking the land must inevitably wear it away. 'There is not a torrent, nor any flow of water, however small, but carries away earth, with gravel or sand to lower levels.... We never see a river in flood, but we must acknowledge the carrying-away.... of our land, to be sunk at the bottom of the sea; we never see a storm upon the coast, but we are informed of a hostile attack of the sea upon our country'. Once it enters the sea, sedimentary detritus, as the fragmental material resulting from the weathering and erosion of solid rocks is known, is 'carried farther and farther along the shelving bottom of the sea.... by the agitation of the winds, tides and currents'. It eventually comes to rest as loose and incoherent sediment, deposited in flat-lying layers upon the sea-floor. Only later is it converted into solid rock.

But then Hutton argued that, given sufficient time, and 'in nature, we find no deficiency with respect to time', weathering and erosion must inevitably reduce the land-surface until it was covered by the sea, so destroying any basis for plant and animal life, at least upon the land. It was a conclusion that Hutton viewed with equanimity, reasoning if 'the constitution of this world had been wisely contrived', there would be some process of consolidation and upheaval whereby 'the necessary decay is naturally repaired', so forming new and habitable land from the ruins of a more ancient world. Otherwise, 'the system of this earth has.... been intentionally made imperfect' or 'it has not been the work of infinite power and wisdom'. Hutton evidently viewed such propositions as anathema.

Central to Hutton's argument was the observation, made many years earlier, that a 'vast proportion of the present rocks [exposed upon the land] is composed of materials afforded by the destruction.... of more ancient formations'. Moreover, Hutton stressed that limestones were often present within these ancient formations, so providing him with incontrovertible evidence for the existence of habitable worlds in the distant past, since were they not composed of the remains of once-living organisms? But if these sedimentary formations had been deposited upon the sea-floor, they must have subsequently been raised above sea-level to form the foundations of the present world as it now exists, especially as they no longer preserved the flat-lying attitudes of their original deposition. The alternative explanation, that the sea-level had fallen, Hutton rejected on the grounds that it was impossible to say where the waters had gone.

Even before he left Sligh Houses, Hutton clearly realised not only did the world we now inhabit arise from the waste of a former world, but equally there was no logical reason to suppose that such a former world had not itself arisen from the ruins of an even older world. But then the world we now inhabit must be just as vulnerable to the very same processes of destruction. Accordingly, the sedimentary rocks now being deposited upon the sea-floor would surely be raised up to form new lands, once the present-day continents had finally been worn away by weathering and erosion. Indeed, there was no logical reason why such cycles of destruction and renewal in producing a 'succession of former worlds' should not be repeated indefinitely into the future, just as they had occurred in the past.

He concluded: 'The result, therefore, of our present enquiry is that we find no vestige of a beginning, no prospect of an end'.

Principle of Uniformitarianism

But Hutton could only reach such a conclusion by arguing that 'the present is the key to the past', at least as far as the geological processes moulding the earth's surface features were concerned. He wrote: 'no powers are to be employed that are not natural to the globe, no action to be admitted.... except we know the principle, and no extraordinary events to be alleged in order to explain a common appearance. The powers of nature are not to be employed in order to destroy the very object of those powers, we are not to make nature act in violation to the order we actually observe.... Chaos and confusion are not to be introduced into the order of nature, because certain things appear to our practical views as being in some disorder.... We are not to suppose that there is any violent exertion of power, such as is required in order to produce a great event in little time. In nature, we find no deficiency with respect to time, nor any limitation with regard to power'.

Such a philosophical argument would later be promulgated as the doctrine of uniformitarianism. It supposes that geological processes have remained much the same throughout geological history. Even if these processes may not always have operated in the distant past at exactly the same rate as at the present day, they were nevertheless governed by the same laws of nature as we presently understand them. Hutton's greatest contribution to geology thus rests in realising that the earth's surface has undergone a series of revolutionary changes in response to everyday processes operating over vast periods of geological time.

'The Abyss of Time'

Yet Hutton's *Theory of the Earth* could only be verified in the field. Quite by chance, while visiting a friend near Jedburgh late in 1787, he first observed what later became known to geology as an **angular unconformity**. Exposed in the banks of the River Jed were flat-lying sandstones and shales, red in colour, resting with an abrupt break upon vertical beds of what would later be known as greywackes and shales, but which Hutton called *schistus*. Separating the two sequences was a thin bed of puddingstone, or conglomerate, made up of rounded fragments of the underlying rocks, clearly derived from their weathering and erosion.

Then in June 1788, searching for further evidence to corroborate this dramatic discovery, James Hutton, accompanied by Sir James Hall of Dunglass, and John Playfair, sailed along the Berwickshire coast to reach the now-famous locality of Siccar Point, three miles south of Cockburnspath. There, in Hutton's words, 'we found the junction of that schistus [the older formations of greywacke and shale] with the red sandstone... at Siccar Point.... washed bare by the sea. The sandstone strata.... are partly remaining upon the ends of the vertical schistus; and in many places, points of the schistus are seen standing up through among the sandstone.... Behind this again we have a natural section of these sandstone strata, containing fragments of the schistus.... We returned perfectly satisfied'.

The discovery made a profound impression on Hutton's companions. As Playfair later wrote: 'The palpable evidence presented to us.... gave a reality and substance to those theoretical speculations, which, however probable, had never till now been directly authenticated by the testimony of the senses. We often said to ourselves, what clearer evidence could we have had of the different [ages of] formation of these rocks, and of the long interval which separated their formation, had we actually seen them emerging from the deep? We felt ourselves necessarily carried back to the time when the vertical schistus on which we stood was yet at the bottom of the sea, and when the sandstone before us was only beginning to be deposited, in the shape of sand or mud, from the waters of a superincumbent ocean. An epoch still more remote presented itself, when even the most ancient of these rocks, instead of standing upright in vertical beds, lay in horizontal planes at the bottom of the sea, and was not yet disturbed by that immeasurable force which has burst asunder the solid pavement of the globe. Revolutions still more remote appeared in the distance of this extraordinary perspective. The mind seemed to grow giddy by looking so far into the abyss of time.... and we became sensible of how much farther reason may sometimes go than imagination may venture to follow'.

Reception of Hutton's Theory

Hutton's *Theory of the Earth* was bitterly criticised after its initial publication in 1788. Not only was he charged with atheism, but his scientific views were ridiculed, especially by the disciples of Abraham Gottlieb Werner (1750-1817), Professor of Mining and Mineralogy at Frieberg, who later became known as the Neptunists.

Admittedly, Hutton laid himself open to criticism when he appealed to the effects of subterranean heat in accounting for the consolidation of sedimentary rocks as a central tenet of his theory. He supposed it acted by fusing their constituent particles together, so converting sediment deposited as loose and incoherent material upon the sea-floor into solid rock, perhaps aided by the weight of the overlying ocean.

But such a view ran counter to the conventional wisdom of the time that sedimentary rocks could be cemented together by mineral matter deposited in the pore-spaces between the sedimentary grains by groundwater, while fossils could be petrified by stony or metallic material replacing their organic matter. Hutton argued that the groundwater could not escape, but it is now realised that compaction occurs in response to the weight of the overlying rocks, driving out any pore-water in the sediment, even as its grains become cemented together by mineral matter in what is a complex series of chemical changes.

Yet subterranean heat was still fundamental to Hutton's *Theory of the Earth*, since it provided the driving force whereby sedimentary rocks were raised from the depths to form new and habitable land. So elevated, they are often found 'broken, twisted, and confounded, as might be expected from the operation of subterranean heat and violent expansion'. Not only are sedimentary strata, originally deposited as flat-lying layers upon the sea-floor, now often found tilted at varying angles away from the horizontal, becoming highly inclined or even vertical in some cases, but locally they often display folds and faults, wherever sedimentary strata are contorted or disrupted by earth-movements. *Folds* are simply flexures in sedimentary strata, bent without any break into what Hutton variously called inflexions or sinuosities, while *faults* are actual breaks in sedimentary strata, across which movement has taken place, the rocks on either side slipping or shifting past one another.

Neptunist Theory of Werner

Quite a different explanation was given by Werner to such distortions in sedimentary strata within the context of his own 'theory of the earth'. He argued for an universal ocean, chaotic in nature, encircling what was originally the nucleus of the primeval earth. The waters of this universal ocean were remarkable in containing all the mineral matter needed to deposit the rocks of the earth's crust. It was upon the highly irregular floor of this universal ocean that chemical precipitation first occurred, forming what Werner called the 'primitive' rocks. Crystalline in nature, he thought that such rocks could only be precipitated from what was an aqueous solution, however hot and concentrated. Werner included a coarse-grained rock known as granite as among the oldest of these 'Primitive' rocks.

Then, after their depostion, the waters of the universal ocean started to ebb away, revealing the topmost peaks of what would eventually become mountains. It was upon the lower slopes of these submarine mountains, still partly covered by the waters of the universal ocean, that the 'transitional' rocks were deposited at varying angles to the horizontal. Overlying the 'primitive' rocks, they included not only chemical precipitates, giving rise to rocks with crystalline textures, but also mechanical accumulations of sedimentary detritus, such as greywackes and shales, derived from what was now dry land, exposed to weathering and erosion. They were overlain in turn by 'secondary' rocks, lying closer to the horizontal, and consisting largely of sedimentary rocks such as sandstones, limestones and shales.

Bedding of Sedimentary Rocks

Werner did not believe that the 'transitional' rocks in particular were originally deposited as flat-lying layers. It was therefore reasonable enough to account for the folding, and other signs of disturbance shown by such rocks, by supposing they had slumped under gravity down the submarine slopes on which they were originally deposited. To rebut such a hypothesis, Hutton observed that *sedimentary rocks*, whether or not they are folded, nearly always occur in distinct beds, each maintaining an almost constant thickness over wide areas. Indeed, such a feature is characteristic of the *bedding* or *stratification* shown by sedimentary rocks.

He argued that such a characteristic feature of sedimentary rocks is only compatible with their deposition as flat-lying layers, close to the horizontal. Transported back-and-forth across the sea-floor by the combined action of waves, tides and other currents, any loose and incoherent sediment must inevitably fill in any irregularities, so forming a surface very close to the horizontal. But exactly the same processes operate once another bed of sedimentary rock is then deposited on top of this surface. Such deposition must form another surface parallel to the first, since it too will be close to the horizontal.

Hutton thus accounted for the parallel nature of sedimentary bedding by supposing that sedimentary beds were nearly always deposited as horizontal layers, quite contrary to Werner's arguments. This was not to say that deposition could not occur upon a sloping surface, but the sedimentary beds deposited in such circumstances would no longer be found lying almost exactly parallel to one another. In fact, they are then said to display an *initial dip*, while they typically occur as wedge-shaped beds, thickening towards their source.

Origin of Basalt and Granite

Even so, it was the explanation that Werner gave for rocks now known to be volcanic in origin that caused his whole system to founder. He had included a hard, dark, compact, and fine-grained rock known as *basalt* in his category of 'secondary' rocks, implying that it was a chemical precipitate, deposited from the still-receding waters of his universal ocean. However, Nicholas Desmarest (1725-1815) had already recognised basalt as a typical lava with gas-bubbles and columnar jointing, erupted from the extinct volcanoes of the Auvergne in France. The Vulcanists, as the advocates of this theory became known, applied the volcanic origin of basalt, not only to lavas recently and quite obviously erupted from volcanoes, but also to its every occurrence, interbedded with sedimentary rocks.

Coming early to this view around 1760, Hutton later emphasised that basalt was not only erupted as molten lava from volcanoes at the earth's surface, but he argued that it could also have 'flowed by heat among the strata of the globe' to form what he called 'subterranean lavas'. Occurring respectively as lava-flows and igneous intrusions, basalt is thus merely one variety of igneous rock, formed from molten rock-material or *magma* as it cools and solidifies, often forming a crystalline rock rather than a volcanic glass.

While Hutton realised all the implications of the Vulcanist theory as it applied to basalt, he was the first observer to view the origin of granite in a similar light. *Granite* is a crystalline rock, coarse-grained and light in colour. It typically underlies wide tracts of the countryside, forming large subterranean masses, especially in mountainous regions. Werner argued that it only occurred as the earliest-known precipitate from the waters of his universal ocean, so that it was the oldest of all rocks. But Hutton, suspecting its intrusive nature, undertook several field-excursions within Scotland, trying to verify its igneous origin.

Surprisingly, his very first journey in 1785 yielded incontrovertible evidence from Glen Tilt, where he observed narrow veins of granite in the bed of the River Tilt, cutting across the stratification of the surrounding marble. The field relations clearly demonstrated that granite could not be the 'original or primitive part of the earth'. Instead, such veins of granite could only be younger than the rocks they intrude. Their intrusive nature was best explained by supposing that the granite had penetrated the marble as veins of molten rock, which then crystallised out from a molten state to form a solid rock. Subsequent excursions to Galloway and Arran merely served to confirm Hutton's findings from Glen Tilt.

Once granite had been identified as igneous in origin, there were no rocks which could be recognised as 'primitive', in the sense used by Werner, implying that they were the first rocks ever to be formed. Instead, granite masses were inevitably found to be surrounded by older country-rocks, which they intruded. These country-rocks often displayed some evidence of a sedimentary origin, as Hutton recognised, but they are often so altered that they later came to be distinguished as *metamorphic rocks*, quite separate from the sedimentary and igneous rocks from which they were formed. Typically, metamorphic rocks are crystalline in nature, displaying the effects of temperature, pressure, and earth-movements, acting deep within the earth's crust upon pre-existing rocks. This caused the growth of new minerals within the solid rock, while their structural features were profoundly changed by the effects of deformation and recrystallisation.

Rise of Historical Geology

Hutton was content merely to accept Werner's simple division of what became known as the 'primary' and 'secondary' rocks, along with the intervening category of 'transitional' rocks, and indeed he made no attempt to quantify geological time. Instead, it was Werner, and his disciples on the Continent, who first came to appreciate that sedimentary rocks were superimposed upon one another in a definite order. Indeed, they occurred as distinct formations like the Carboniferous Limestone or the New Red Sandstone, at least within a particular region. Then, early in the 19th century, it came to be realised that the stratigraphic age of such formations could also be established by studying the fossils found within them.

Indeed, it was the sudden appearance of hard-bodied animals in sedimentary rocks that allowed the shells or other hard parts of these animals to be preserved as fossils. This occurred at the start of Cambrian times some 570 million years ago, allowing geological time to be divided into two distinct eras. The Precambrian era lasted from the formation of the earliest known rocks around 4,000 million years ago to the start of Cambrian times. There are very few fossils preserved in rocks of Precambrian age, apart from very small organisms such as plant spores and the like, along with the impressions of soft-bodied animals, making it difficult to divide up Precambrian time, except in a very broad manner.

However, the subsequent evolution of hard-bodied animals at the start of Cambrian times allowed a much more detailed time-scale to be devised for the more recent events in the geological history of the Earth. The subsequent evolution of particular species and genera during the course of this history, coupled with their eventual extinction, means that rocks can be dated geologically by studying the fossils found within them. The nature of this record is such that this latter part of geological time can be divided up into a number of different periods. Each is given a name, and taken together, these various periods of geological time then constitute what is known as the geological timescale, as shown in the stratigraphic column (see Table on page vi). The ages in millions of years given for these geological periods can be determined by analytical techniques, based on the breakdown of radioactive minerals in a rock.

CHAPTER 2

Nature of the Geological Record

GEOLOGY HAD THUS EVOLVED by the early decades of the 19th century into essentially an historical science, much concerned with the dating of rocks. Partly, this could be achieved, as Hutton had so admirably demonstrated in undertaking his excursions to Glen Tilt and Siccar Point, simply by observing their relationships to one another in the field. But equally important, sedimentary rocks contain fossils as the remains of once-living organisms, which give a measure of their stratigraphic age. Yet this is only one aspect of the geological record. Rocks are equally the product of physical processes acting during the course of geological history in accordance with the Principle of Uniformitarianism. Thus they can be divided according to their mode of origin into three distinct categories: sedimentary, igneous and metamorphic.

Sedimentary rocks are derived from the weathering and erosion of pre-existing rocks. This produces sedimentary detritus, chiefly mineral grains and rock fragments, which is mostly carried across the Earth's surface by water, only to accumulate away from its source as flat-lying layers of sedimentary rock.

Igneous rocks are formed wherever molten lava, or its underground equivalent, cools and solidifies. If this occurs very rapidly, molten lava may be quenched so that it forms a glassy rock, lacking any crystalline structure. More commonly, minerals start to crystallise out from such a melt, and when this crystallisation is complete, a solid rock is formed.

Metamorphic rocks are formed wherever pre-existing rocks are so altered in response to extremes of temperature and pressure that they take on an entirely different character. These changes occur in the solid rock as a result of chemical reactions between its original minerals, causing these minerals to recrystallise, or allowing new minerals to grow in their place.

Nature of Sedimentary Rocks

Sedimentary rocks are composed of material derived from the weathering and erosion of pre-existing rocks. They are therefore a major repository of geological history, reflecting the nature of the environmental conditions under which they were originally deposited as flay-lying layers upon the earth's surface. As already mentioned, *bedding* or *stratification* is a characteristic feature of sedimentary rocks, which typically occur as *sedimentary beds*, separated from one another by *bedding-planes*. Rarely more than a few feet thick, sedimentary beds can often be traced for considerable distances in the field. Individual beds can usually be distinguished in the field by differences in grain-size, texture, colour, hardness and mineral composition, which are often accentuated by the effects of weathering and erosion. Sedimentary rocks are deposited in chronological order on top of one another to form a *sedimentary sequence*. The oldest beds always occur occur at its base, passing upwards into ever younger horizons towards its top, according to the *Principle of Superposition*. It allows sedimentary rocks to be dated with respect to one another, even if they contain no fossils.

Conglomerates and Breccias. These rocks are just accumulations of rock fragments, evidently derived by the weathering and erosion of older rocks, set in a much finer-grained matrix. *Conglomerates* consist of pebbles and boulders, which have probably been transported for some distance, to judge by their rounded nature. *Breccias* consist of angular rock fragments without any rounding of their corners, obviously derived from a nearby source.

Sandstones, Arkoses and Greywackes. Weathering and erosion may also break down the mineral grains in pre-existing rocks into much smaller particles, so providing another source of sedimentary material. Chemical breakdown typically leads to the formation of clay particles, microscopic in size, while leaving quartz and other resistant minerals unaffected. The mineral grains which escape the effects of chemical breakdown are then carried away from their source by running water, eventually to be deposited as sand.

Sandstone is formed wherever these mineral grains become compacted and cemented together to form a solid rock. Most sandstones consist of quartz grains, since this mineral is particularly resistant to chemical weathering. However, the physical breakdown of rocks like granite and gneiss under conditions of very rapid erosion often leads to the presence of much feldspar as well as quartz in the rock, which is then known as *arkose*. Another type of sandstone consists of a wide variety of rock fragments and mineral grains, set in a finer-grained matrix rich in clay minerals. Known as a *greywacké*, it is not a common rock in the Scottish Highlands.

Quartzites. The sedimentary grains in a sandstone can usually be seen through a hand lens of moderate magnification on a weathered surface, particularly if they are cemented together by calcite, dolomite or hydrous iron oxides. However, where quartz acts the cement, the detrital grains become much less conspicuous, unless feldspar is present as well as quartz in the rock. If such a sandstone forms a very hard rock, which typically breaks across the mineral grains, it is known as a *quartzite*. Such rocks are usually white or grey in colour.

Cross-bedding and Graded Bedding. Sandstones laid down in relatively shallow water display *cross-bedding* wherever a sedimentary bed is cut by internal bedding-planes, inclined at an angle to the upper or lower contacts of the bed itself. Often, these internal bedding-planes are truncated by erosion along the upper contact, while they become flat-lying as they are traced towards the lower contact. *Graded-bedding* occurs wherever the grain-size decreases towards the top of a sedimentary bed, away from its base. This is rather more typical of sandstones like greywackés, which were deposited in relatively deep water. Cross-bedding and graded-bedding can both be used to determine if sedimentary beds are right-way-up or upside down.

Shales and Mudstones. Carried away from their source by running water, the clay particles produced by the chemical breakdown of pre-existing rocks are often transported far and wide. Eventually, they settle out of suspension in still water, most usually as mud on the sea-floor. Such deposits first form clays, and then harden into shale and mudstone. *Shale* is a very fine-grained sedimentary rocks composed of clay minerals, which splits easily along the bedding, unlike *mudstone*, which is not so fissile. Both are susceptible to further weathering and erosion, so that they are often not well-exposed at the surface.

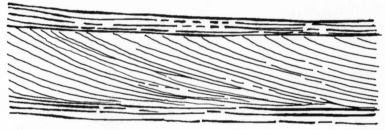

Figure I. Cross-bedding in Sandstone

Limestones and Dolomites. Chemical weathering also produces material in solution, which eventually finds its way into the oceans. There, concentrated by evaporation, it forms salt water. These salts, dissolved in sea-water, can be precipitated directly as a result of further evaporation, so forming various types of salt deposit, known as *evaporites*, or they can be abstracted by living organisms to form their hard parts. When these animals die, their hard parts may then accumulate as sedimentary deposits on the sea-floor.

Limestones in particular are formed by a combination of these processes, being partly chemical precipitates and partly the accumulation of organic remains. They are calcareous rocks rich in calcium carbonate, which occurs in the form of calcite, and not surprisingly, they often have fossils preserved within them. Once deposited, limestones may be altered chemically by brines rich in dissolved salts to form *dolomite*. This rock is largely composed by the mineral also known as dolomite, which is a carbonate of magnesium as well as calcium. Calcite and dolomite are easily scratched with a knife, and this serves to distinguish these minerals from the quartz in sandstones.

Limestones and dolomites are rocks easily dissolved by rain water, carrying carbon dioxide in solution. A landscape typical of limestone is then found with flat-lying outcrops occurring as limestone pavements. Where rain-water runs down any steep faces on the limestone, it often dissolves away the rock to make an intricately fretted surface with very sharp edges. Typically, limestone often makes fertile ground with good soil, covered in grass, unlike sandstone and especially quartzite, which produce acid soils, often only covered by heather or bracken.

Evidence for Earth-Movements

The sedimentary record suggests that movements of the earth's crust have occurred throughout the course of geological time. Apart from limestones and dolomites, and evaporite deposits, sedimentary rocks can only be deposited if there is a suitable source of detrital material. Such a supply can only be maintained if uplift exposes ever more rocks to weathering and erosion within the source area. Likewise, there is only space for the long-continued accumulation of sedimentary rocks if the earth's crust continues to subside, so forming a sedimentary basin.

Dip and Strike. Wherever earth-movements have affected sedimentary rocks after their deposition, their bedding is found to be tilted at an angle, away from the horizontal. A compass can first be used to find the *strike* of the bedding. By drawing a horizontal line across a bedding-plane, its strike can be measured as the bearing of this line from true north. A clinometer is then used to measure the angle made by the bedding plane to the horizontal. This gives the *dip* of the bedding plane at right angles to the strike. The direction of dip also needs to be recorded.

Anticlines and Synclines. Any changes in dip often reflect the presence of folds in sedimentary rocks, whereby the bedding is bent into waves-like shapes. Typically, *anticlines* (arches or up-folds) are separated from one another by *synclines* (troughs or down-folds). The bedding then dips away from the *hinges* of the anticlines towards the synclines, unless the folds are overturned, so that both the *fold-limbs* dip at different angles in the same direction.

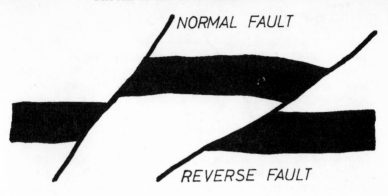

Figure 2. Normal and Reverse Faults.

Joints and Faults. Earth-movements can also cause rocks to fracture and break, forming joints and faults. *Joints* are seen in virtually every exposure, forming closely-spaced fractures in the rock, across which no movement can be detected. They often occur in a very regular manner, forming one or more sets of parallel fractures, spaced at a regular interval, particularly in sedimentary rocks. *Faults* differ from joints in that actual movement has occurred along the fracture. This results in a *fault-plane*, across which the rocks on either side have slipped or shifted relative to one another, parallel to the fracture itself.

Normal faults occur wherever the fault-plane itself dips in the direction of its downthrow, often at a rather steep angle. Exactly the opposite relationship is seen in a *reverse fault*, where the overlying rocks have been thrust up the fault-plane, which often dips at rather a shallow angle in the opposite direction. Horizontal movements can also occur on a fault-plane, giving rise to a *strike-slip fault*. The Great Glen Fault is a good example, formed by the Northern Highlands moving southwest past the Grampian Highlands over a distance of some 40 miles.

Angular Unconformities. As Hutton first demonstrated, angular unconformities are produced in response to earth-movements, which often reach their climax in the folding and faulting of sedimentary rocks. These rocks then undergo uplift, often associated with mountain-building, so that they become exposed to weathering and erosion above sea-level. Such processes typically produce a surface of erosion, cutting across the bedding and other structures of these older rocks. If deposition were then renewed, a younger sequence of flat-lying sediments would be laid down unconformably across these older rocks.

Such an angular unconformity thus represents a break in deposition, marked by uplift and erosion at a time when earth-movements had also affected the underlying rocks. Its essential feature is the presence of a structural discordance between an older group of rocks, whether sedimentary, igneous or metamorphic, and a younger sequence of sedimentary rocks. Angular unconformities typically occur at different stratigraphic levels, separating sedimentary sequences of different age from one another. Geological history therefore has a cyclical nature, whereby sedimentary rocks once deposited are folded and faulted, and then undergo uplift and erosion, only to be buried by renewed deposition as subsidence takes the place of uplift.

Surfaces of Erosion. Angular unconformities often represent surfaces of erosion that were themselves very close to the horizontal when the overlying rocks were deposited. This typically occurs where the sea invades the land, reducing the old land-surface to a flat-lying plane of marine erosion, which then serves as the foundation for more sedimentary rocks to be deposited. If these rocks were laid down as beds very close to the horizontal, as commonly the case, the bedding of these rocks would then be parallel to the unconformity.

Figure 3. Angular Unconformity and a Buried Landscape.

However, erosion might not always be able to reduce the surface of unconformity to a horizontal plane before deposition was renewed. This typically occurs where the unconformity represents an ancient landscape, which still preserves a degree of topographic relief. The younger sediments are then often found banked against the older rocks, so displaying an initial dip, while wedging out against these older rocks along the unconformity itself. Further deposition then results in a *buried landscape*, which can be exhumed at a later date.

Nature of Igneous Rocks

Deep within the earth, temperatures are sufficiently high for melting to occur, at least locally. The molten rock so formed is known as *magma*, and it eventually solidifies to form the *igneous rocks*. If magma escapes towards the surface, it often erupts as lava from volcanoes, cooling down and solidifying as *lava-flows*. They typically erupt one after another, so forming a volcanic sequence in much the same way as sedimentary rocks accumulate as flat-lying layers.

However, if volcanic eruptions are more explosive in nature, the force of these explosions may be sufficient to shatter the walls of the volcanic vent, producing much fragmental material. *Pyroclastic rocks* then accumulate within and around the vent, forming *agglomerate* if the fragments are so large that they can easily be seen, or *volcanic ash* or *tuff* if it is much finer in grain-size.

Igneous Intrusions. Magma in forcing its way upwards usually encounters colder country-rocks as it rises towards the surface. Often, it cools down and solidifies while still at considerable depths within the Earth to form *igneous intrusions*. As they are surrounded on all sides by older rocks, they can be dated as younger than their *country-rocks*. The heat carried by igneous intrusions often alters these country-rocks, forming what are known as *metamorphic aureoles*.

Some intrusions of igneous rock occupy vertical fractures which the magma wedged apart as it flowed upwards, so forming a wall-like mass known as a *dyke*. Other intrusions are the result of magma flowing horizontally along the bedding of sedimentary rocks, lifting up its roof to form a sheet-like body known as a *sill*. Both dykes and sills often show *columnar jointing*, where joints are formed at right angles to the surfaces against which the magma cooled. Igneous intrusions also occur as very large masses of igneous rock, known as *batholiths*, which appear to lack any foundation of older rocks at depth.

Textures of Igneous Rocks. Wherever lava erupts from volcanoes, it tends to cool down very quickly, occasionally passing into a glassy state, without any crystallisation taking place at all. More usually, however, a fine-grained rock with a crystalline texture is produced, typical of lava-flows. Unless the lava carried up larger crystals from depth, so forming *phenocrysts*, it is not easy to see the individual grains in the rock, even with a hand-lens. Igneous rocks consisting of phenocrysts set in a much finer-grained matrix are said to display a *porphyritic texture*.

Dykes and sills resemble lava-flows in that they cool quickly in contact with their country-rocks, so that they usually consist of relatively fine-grained rocks as well. They often exhibit *chilled margins* against their country-rocks, marked by narrow selvedges of much finer-grained or even glassy rock. Batholiths and other deep-seated igneous intrusions invariably cool down very slowly, perhaps taking several million years to reach the temperature of their surroundings. This only allows a relatively small number of largish crystals to grow as the magma solidifies, rather than a multitude of much smaller ones, so producing a coarse-grained rock as a result.

Rock-forming Minerals. Nearly all rocks are made up of different minerals, occurring together as crystalline grains in the rock. Apart from calcite and dolomite, which are carbonate minerals, silicates are the most common rock-forming minerals, since the earth is mostly composed of silicon and oxygen. These elements combine in various ways with aluminium, sodium, potassium, calcium, iron and magnesium to form a wide variety of silicate minerals, often with very complex compositions.

Among the most important of the silicate minerals are the *feldspars*, which are complex alumino-silicates of sodium, potassium and calcium. Feldspar occurs as pale pink or whitish grains, typically showing planes of cleavage, parallel to one another, across which the individual crystals split apart. It is difficult to scratch with a knife. Where there is not sufficient sodium, potassium or calcium in a rock to combine with silicon and oxygen to form feldspar, silica SiO_2 may be found in the form of *quartz*. Lacking any cleavage, this colourless mineral usually has a "glassy" appearance. Vein quartz often has a milky-white colour. Quartz is slightly harder than feldspar.

The other silicate minerals commonly found in igneous rocks are often dark in colour, owing to the presence of iron, as well as magnesium. *Olivine* is a relatively simple silicate of iron and magnesium, often greenish in colour as the name suggests. *Pyroxenes* such as augite are much more complex silicates of calcium, iron and magnesium, often combined with some aluminium. *Amphiboles* such as hornblende are rather similar, except that they are hydrous silicates of calcium, iron and magnesium, often with sodium and aluminium as well. *Micas* are another group of silicate minerals commonly found in igneous rocks, occurring as tabular crystals with a perfect cleavage in one direction. *Biotite* is a dark mica, rich in magnesium and iron, while *muscovite* is a white mica, rich in aluminium.

Naming Igneous Rocks. Igneous rocks were first classified according to the amount of silica SiO_2 in the chemical analysis. *Acid rocks* like granite are rich in silica, potash and soda, but poor in iron oxides, magnesia and lime, so that they are usually light in colour. *Basic rocks* such as gabbro are the reverse, rich in iron oxides, magnesia and lime, but poor in silica, potash and soda, so that they are typically dark in colour. Igneous rocks lacking any feldspar are known as the ultrabasic rocks.

Granite is a light-coloured rock composed mostly of potash feldspar, possibly with soda feldspar as well, together with an appreciable amount of quartz. Hornblende and biotite are common as accessory minerals, giving a somewhat speckled appearance to the rock. Granite may also occur as *pegmatite*, which is an extremely coarse-grained rock consisting of quartz and feldspar, together with muscovite in many cases.

Gabbro is a much darker rock than granite, consisting mostly of calcic feldspar and pyroxene in roughly equal amounts. Olivine is often present in the rock. There is, however, a whole range of igneous rocks, intermediate in composition between granite and gabbro, such as granodiorite, quartz- diorite and diorite. These rocks are often just called granite in the field, all sharing much the same texture.

Coarse-grained igneous rocks have their fine-grained equivalents in the volcanic rocks, typically occurring as lava-flows, or dykes and sills, rather than deep-seated intrusions. The fine-grained equivalent of granite is *rhyolite*, which mostly occurs as a pale flow-banded rock. *Andesite* is a much more basic rock, closer in chemical composition to gabbro, often forming a rather dark rock of purple hue. *Basalt* is the fine-grained equivalent of gabbro. It occurs as a very dark or even blackish rock, which weathers easily to form a brownish coating on its surface. *Dolerite* is the intrusive and somewhat coarser- grained equivalent of basalt, often found making up fairly large sills.

Nature of Metamorphic Rocks

Metamorphism occurs wherever pre-existing rocks buried deep within the earth are so altered that they take on an entirely different character, even though all these changes occur while the rock still remains solid. Such changes often occur in response to the influx of heat, carried perhaps by igneous intrusions. This causes the pre-existing minerals in the rock to recrystallise, often forming a coarser-grained rock as a result. Entirely new minerals may also be formed in the solid rock, so that metamorphism often changes both the texture and the mineral composition of the original rock. An aureole of *contact* or *thermal metamorphism* is produced around an igneous intrusion where such changes affect the nearby country-rocks.

However, metamorphism is often much more pervasive in its effects, particularly where very large masses of sedimentary and igneous rocks are affected by *mountain-building*. It is often accompanied by intense deformation which completely alters the structural features shown by the metamorphosed rocks, as well as altering their mineral composition. Since these changes typically affect rocks exposed over very wide areas, they come under the heading of **regional metamorphism**.

Naming Metamorphic Rocks. The nomenclature of metamorphic rocks is relatively simple, since it is based essentially on their textural and structural features, as developed by metamorphism in various types of sedimentary and igneous rocks. Sandstones and limestones often just recrystallise to form *quartzites* and *marbles*, without any change in their mineral composition. They are usually coarser in grain than the sandstones and limestones from which they were formed. Such rocks are found in aureoles of contact metamorphism, or as a product of regional metamorphism.

Hornfels. Shales and mudstones are much more susceptible to changes in mineral composition. Contact metamorphism typically converts them into a hard and very splintery rock known as a *hornfels*. This rock is tough and rather fine-grained, often purplish in colour. Any pre-existing structures such as bedding are often obscured or even completely destroyed by the effects of recrystallisation and the growth of new minerals in the solid rock.

Slates, Phyllites and Schists. Quite different rocks are produced from shales and mudstones as a result of regional metamorphism. Where intense deformation affects these rocks at relatively low temperatures, they are converted into *slates*. A *slaty cleavage* is imposed on the rock, whereby it splits into extremely thin sheets, lying oblique to the bedding. It is produced by recrystallisation of the original clay minerals in the rock so that they all lie parallel to one another. Slates pass into *phyllites* as the rock becomes slightly coarser in grain, imparting a silky sheen to the cleavage surfaces.

Under conditions of higher temperature, and possibly in response to greater pressures, slates and phyllites pass in their turn into *schists*. These are distinctly crystalline rocks, in which micas like biotite and muscovite all lie roughly parallel to one another, forming a *schistosity*. Often, there are other metamorphic minerals present in the rock, such as garnet, so forming *garnetiferous mica-schists*. Igneous rocks like basalt and dolerite may also be converted into *hornblende-schists* or *amphibolites* under similar metamorphic conditions.

Gneisses. Even coarser-grained rocks may be produced in response to regional metamorphism, giving rise to *gneisses*. They are more even-grained than schists, partly owing to the presence of abundant feldspar in the rock. Lacking a schistosity, they often occur as banded rocks with distinct layers, differing in mineral composition from one another. Gneisses can be formed from a very wide variety of original rocks, in response to the high temperatures and great pressures which exist deep within the earth's crust. Gneisses typically occur as very old rocks, which are often found underlying all the other rocks of a particular region, forming its geological foundations. Such rocks then constitute what is known as a *basement complex*.

Mylonites. Metamorphism usually results in more coarse-grained rocks than originally the case, even if recrystallisation and the growth of new minerals are accompanied by deformation. However, the deformation may be so intense that the original minerals in the rock are broken down into much finer-grained material, particularly where low temperatures do not favour much recrystallisation. The rocks formed as a result are known as *mylonites*. They are extremely fine-grained rocks with a banded or streaky appearance in the field.

All these rock-types are illustrated in *A Photographic Guide to Minerals, Rocks and Fossils*, (New Holland, 1998), with text and photographs by the author.

CHAPTER 3

Geological Record in the Scottish Highlands

THE SCOTTISH HIGHLANDS ARE underlain by rocks which differ widely in age. Very ancient rocks are found in the Northwest Highlands and Outer Hebrides, where the *Lewisian Gneiss* is exposed over wide areas. It is overlain unconformably by the much younger rocks of the *Torridonian Sandstone*, followed in its turn by a sequence of *Cambro-Ordovician* rocks. All these rocks are separated from the rest of the Scottish Highlands by a great dislocation, known as the *Moine Thrust*. It marks the western limit to the Caledonian (and earlier) earth-movements, which affected the metamorphic rocks now forming the rest of the Scottish Highlands, towards the end of Silurian times, around 400 million years ago.

These metamorphic rocks are known collectively as the *Highland Schists*. They fall into two distinct groups. The *Moine Schists* are the older rocks, and they have had a very complex history. Bounded to the northwest by the Moine Thrust, they are found throughout much of the Northern Highlands and beyond the Great Glen in the Grampian Highlands. Intensely deformed areas of Lewisian Gneiss are also found in the Northern Highlands, entirely surrounded by Moine Schists. The younger group of metamorphic rocks is known as the *Dalradian Series*. It outcrops over a wide area in the Grampian Highlands, southeast of the Moine Schists. Both Moine and Dalradian rocks are intruded by a great many igneous masses, mostly of granite and related rock-types.

Even younger rocks are exposed as a sedimentary fringe to the Highlands along the shores of the Moray Firth, and farther north into Caithness and the Orkney Islands. They occur as the Devonian rocks of the *Old Red Sandstone*, overlain locally by *Permo-Triassic* and *Jurassic* rocks. Volcanic rocks of Devonian age are found in Argyll, where they form the *Lorne Lavas*.

Finally, the Inner Hebrides were the focus of much igneous activity in Tertiary times. *Intrusive complexes* forming the roots of ancient volcanoes are now exposed on Skye, Rhum, Mull and Arran, together with Ardnamurchan. *Tertiary lavas* form thick volcanic sequences on Skye and Mull, underlain by Triassic and Jurassic rocks.

Lewisian Gneiss

The oldest rocks found anywhere in the British Isles are formed by the *Lewisian Gneiss*, exposed in the far northwest of the Scottish Highlands, together with the islands of the Outer Hebrides. Indeed, there are few rocks quite as old anywhere else in Europe, apart from some very ancient rocks in Finland, and perhaps even older rocks in the Ukraine. Otherwise, Europe is underlain by much younger rocks, belonging to more recent cycles of geological activity, which separate these ancient massifs from one another.

We have to look much farther west to find such ancient rocks of a similar character, exposed in Greenland and Labrador. Although the Atlantic Ocean now intervenes, this was not always the case. In fact, the Northwest Highlands of Scotland, lying beyond the Moine Thrust, are simply part of a much larger land-mass that was once attached to North America around the end of Precambrian times, some 570 million years ago.

The Lewisian Gneiss forms a *basement complex* of intensely deformed and metamorphosed rocks, cut by igneous intrusions of very different ages, which are themselves often deformed and metamorphosed in their turn. It is not a geological formation in the ordinary sense of the term, as the officers of the Geological Survey first recognised around the turn of this century.

The results of radiometric dating, using the natural breakdown of radioactive minerals in a rock to find its age, now show that the Lewisian Gneiss had a very long and complex history. Dating back to 2,900 million years ago, or even earlier, the evolution of these basement gneisses continued until around 1,400 million years ago. This history therefore represents nearly a third of all geological time since the earth was itself formed, around 4,600 million years ago.

Scourie Dykes. The Lewisian Gneiss often presents the casual observer with a bewildering sense of geological chaos, even within a single exposure. However, it is fortunate for our understanding of this complex that it was intruded by a swarm of igneous dykes, mostly of basic or ultrabasic rocks, around 2,400 million years ago. These intrusions can clearly be seen on the Ten-Mile Map of the Geological Survey, trending northwest-southeast across the Lewisian Gneiss between Loch Laxford and the district of Torridon, much farther to the south. They are known as the *Scourie Dykes*, after the village of the same name, just south of Loch Laxford, where a particularly good example can be seen. Their presence allows the two most important episodes in the evolution of the Lewisian Gneiss to be clearly distinguished from one another.

Thus, where the Scourie Dykes retain their original features, so that they can be recognised as igneous intrusions cutting discordantly across the Lewisian Gneiss, they evidently have not been affected by any subsequent deformation or metamorphism. The gneisses intruded by these dykes must therefore be even older, forming what the Geological Survey originally described as a 'fundamental complex'.

Scourian Gneisses. These very old gneisses are now called Scourian, dating back around 2,700 million years, when they were intensely deformed and metamorphosed under extreme conditions, perhaps corresponding to depths as great as 60 kilometres within the Earth's crust. They occur as *pyroxene-granulites*, consisting of dark and greasy-looking feldspars together with blue opalescent quartz, which gets its milky appearance from the presence of minute needles of rutile within the crystals. Pyroxene rather than hornblende or biotite occurs within these gneisses, as the metamorphism occurred under anhydrous conditions, driving out any water that was once present in these deep-seated rocks.

Laxfordian Gneisses. Where the Scourie Dykes have been so deformed and metamorphosed along with their country-rocks that they can no longer be recognised as discrete intrusions, it is clear that the Lewisian Gneiss has been affected by relatively more recent events in its history. This frequently results in the Scourie Dykes becoming completely incorporated into the surrounding gneisses, so that they now just occur as rather more basic layers within the complex as a whole. The original gneisses also take on an entirely new character, undergoing further deformation and metamorphism under hydrous conditions. These changes typically break down the pyroxenes found in the Scourian Gneisses to form hornblende and biotite.

This later reworking of the 'fundamental complex' results in virtually new rocks, known as *Laxfordian* after the locality of Laxford Bridge. They mostly occur as *hornblende-gneisses*, intruded by much granitic material, often in the form of very coarse-grained pegmatite veins. The results of radiometric dating suggest that the Laxfordian events mostly occurred around 1,800 million years ago, even although some late intrusions give dates around 1,400 million years ago.

Torridonian Sandstone

Resting with a very profound unconformity on top of the Lewisian Gneiss lies the *Torridonian Sandstone*. This abrupt break in the geological record marks a time when the Lewisian Gneiss underwent a vast amount of uplift and erosion, eventually exposing rocks at the surface that were once buried at depths of many kilometres. All this erosion did not reduce these rocks to a flat-lying surface, but it left a landscape of considerable relief, carved into the Lewisian Gneiss. The Torridonian Sandstone was then deposited unconformably on top of these basement rocks, burying this ancient landscape under a flat-lying and very thick accumulation of sedimentary rocks.

Although the Torridonian Sandstone was once thought to form but a single sequence, two distinct groups are now recognised, namely an older Stoer Group and the younger Torridon Group, differing in age by nearly 200 million years. The *Stoer Group* was deposited around 995 million years ago, some 400 million years after the Lewisian Gneiss reached the end of its geological evolution, while the Torridon Group was laid down nearly 200 million years later, around 810 million years ago. Despite this difference in age, the two groups consist of sedimentary rocks, very much like one another.

Stoer Group. The lowermost beds of the Stoer Group are breccias and conglomerates, banked as screes and alluvial fans against hills of Lewisian Gneiss. These rocks pass upwards into red sandstones and mud-stones, probably laid down by braided rivers flowing from much higher ground to the northwest across wide flood-plains. There are also some limestones, most likely deposited in salt-lakes. The Stoer Group was then tilted 30 degrees towards the northwest, prior to the deposition of the Torridon Group, some 810 million years ago.

Torridonian Group. Consisting of a very thick sequence of red sand-stones, rich in detrital feldspar, the Torridon Group can be traced from Cape Wrath in the far north as far as Rhum, lying off the Isle of Skye, some 120 miles to the south. Although the Torridon Group rests uncon-formably on top of the Stoer Group in some places, it is usually found in contact with the Lewisian Gneiss. The unconformity at its base displays only a little relief in the far north, where the underlying gneiss is often weathered to a depth of quite a few feet, forming an ancient soil-profile.

Farther south, however, impressive hills of Lewisian Gneiss are found, occasionally rising to a height of 2,000 feet above their surroundings, and blanketed by the sedimentary rocks of the Torridon Group. Where these rocks have been stripped off the Lewisian Gneiss by recent erosion, this ancient landscape is laid bare, exposed to our view at the present day. Even the valleys once cut into this pre-Torridonian landscape have been exhumed in some places, so that they are now followed by the present-day rivers.

The lowermost sediments of the Torridon Group are very diverse. Coarse breccias occur as scree deposits, banked against the underlying gneiss. They pass upwards into flaggy red sandstones with ripple-marks, together with grey shales showing dessication cracks and ripple-marks. All these rocks are overlain by a very thick and rather monotonous sequence of very coarse-grained and often pebbly sandstones, interbed-ded with more conglomeratic layers. These pebbly sandstones are best described as arkoses, rich in potash feldspar, which are usually dark red or chocolate brown in colour. They were derived from a mountainous area to the northwest as it underwent much physical weathering in response to rapid uplift and erosion, and then deposited by braided rivers flowing away from these mountains.

Cambro-Ordovician Rocks

After the deposition of the Torridon Group, earth-movements again affected the Northwest Highlands, prior to the accumulation of more sedimentary rocks during Cambrian and Ordovician times. After tilting the Torridonian Sandstone away from the horizontal, a long period of erosion then ensued, after which there was a major transgression of the sea across this region at the start of Cambrian times, around 570 million years ago.

Cambrian Quartzite. The lowermost beds of the *Cambro-Ordovician* were then laid down under marine conditions, forming the white and very conspicuous Cambrian Quartzite, which is often seen capping the mountains of Torridonian Sandstone. Typically, this quartzite displays cross-bedding, while there is often a slightly conglomeratic horizon at its very base. The quartzite now dips at 10 or 20 degrees towards the east, while the Torridonian Sandstone below the unconformity is flat-lying.

As the quartzite was once horizontal, this can only mean that the Torridonian Sandstone once dipped at a similar angle towards the west, prior to the deposition of this quartzite. Locally, it cuts across the Torridonian Sandstone, coming to rest directly on top of the Lewisian Gneiss, so forming what has been called the *'double unconformity'* of the Northwest Highlands.

Pipe-Rock. The Cambrian Quartzite passes upwards into an equally quartzitic horizon, known as the *Pipe Rock*. This displays throughout its thickness a multitude of worm burrows, all cutting vertically across the bedding. Some burrows have a simple pipe-like shape, much less than an inch across, while others display more the form of a funnel, so that they are more like a trumpet, opening upwards.

Fucoid Beds and Serpulite Grit. The Pipe Rock is overlain by the *Fucoid Beds*. This distinctive horizon, usually less than 60 feet thick, consists of brown-weathering dolomitic shales, together with some rather more sandy layers, which often show ripple-marks and cross-bedding. It gets its name from the worm burrows commonly seen along the bedding, once identified as the markings made by sea-weed. The *Serpulite Grit* then forms a very persistent horizon of dolomitic quartzite, even thinner than the Fucoid Beds. As the name suggests, some beds are particularly coarse-grained for a sandstone. It carries abundant *Salterella*, which is a very small worm-like organism, originally identified as *Serpulites*.

Durness Limestone. Overlying the Serpulite Grit comes a great thickness of *Durness Limestone*, consisting mostly of dolomite, despite its name. Where fossils are found, they show that the lower part of the Durness Limestone belongs to the Lower Cambrian, while the upper parts are more likely to be Ordovician in age.

Moine Thrust

Separating all these rocks from the rest of the Scottish Highlands is the *Moine Thrust*. It runs up to 20 miles inland along the western seaboard of the Scottish Highlands, from Loch Eriboll and Whiten Head in the north to the Sound of Sleat, where it passes southwest out to sea. Dipping at a very low angle under the rocks lying to the east, it carries the Moine Schists on its back. These rocks have been thrust westwards for tens of miles over the underlying rocks, producing mylonites along the thrust-plane. The thrusting occurred towards the end of Silurian times, marking the very final stages of the Caledonian earth-movements as they affected the Scottish Highlands, some 400 million years ago.

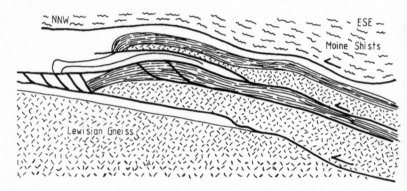

Figure 4. Cross-section through the Moine Thrust.

The rocks underlying the Moine Thrust were also caught up in the movements as the thrusting gradually extended farther and farther north-west. The effect was much like snow piling up in front of a snow-plough. It produced a belt of complex thrusting and folding, up to several miles in width, which affects the rocks now exposed to the northwest, below the Moine Thrust itself. However, these movements died away farther northwest, leaving the Northwest Highlands unaffected by the Caledonian earth-movements. This region now forms what is known as the *foreland* to the Caledonian Mountains, lying to the northwest of the Moine Thrust.

Moine Schists

Lying to the east of the Moine Thrust is a vast area of metamorphic rocks, known as the *Moine Schists*, underlying most of the Northern Highlands and extending across the Great Glen into the Grampian Highlands to the southeast. They display a distinctly monotonous character throughout the whole of their outcrop. Typically, they were once sedimentary rocks, varying in composition from impure sandstones to shales and mudstones, now deformed and metamorphosed as the result of subsequent the earth-movements.

The more shaly rocks have been converted into mica-schists, sometimes rich in garnet and other metamorphic minerals, while the impure sandstones have often just been affected by recrystallisation to form rather massive rocks, rich in quartz and feldspar, together with some mica. Although these rocks are not very schistose, they are quite the commonest rock-type making up the Moine Schists as a whole.

Dalradian Schists

Overlying the Moine Schists of the Grampian Highlands are the *Dalradian Schists*, exposed over a wide area to the southeast. They were originally deposited as a very varied sequence of sedimentary rocks, which was deformed and metamorphosed at the very end of Precambrian times, some 600 million years ago, according to the most recent research. Like the Moine Schists, they display a complex history of folding and deformation, marked by several generations of fold structures.

Quartzites are present at various horizons throughout this sequence, although their place is taken by schistose grits in the Upper Dalradian. Limestones are also a common feature of the sequence, except that they are only found in the Lower and Middle Dalradian. They are metamorphosed to marbles in many places. Shaly rocks originally occurred at virtually all levels within the sequence, but they are now converted into slates, phyllites or schists, according to the intensity of the metamorphism. These rocks sometimes occur as black graphitic-schists, but more commonly as garnet mica-schists.

Boulder-Beds and Pillow Lavas. There are two other very distinctive horizons also present within the Dalradian sequence. Firstly, a *boulder-bed* is found just above the top of the Lower Dalradian, representing the deposits of an ice-sheet, grounded on the floor of a shallow sea. It carries boulders of granite in its upper levels. Secondly, there is a volcanic horizon marking the eruption of *pillow lavas*, just above the top of the Middle Dalradian. Such rocks are formed where molten lava is erupted on the seafloor. As the lava comes into contact with the sea-water, its surface cools rapidly to form a glassy selvedge. This then bursts under pressure, allowing a bulbous protusion of lava to solidify as a pillow-shaped mass of igneous rock.

The underlying Dalradian rocks are also intruded by a great volume of igneous rocks, mostly in the form of basic sills, which are now deformed and metamorphosed along with their country-rocks during an early stage of the Caledonian earth-movements.

Caledonian Granites

The Moine and Dalradian rocks of the Scottish Highlands are intruded by a great many granites, not all of the same age. The earliest granites are found intimately associated with their metamorphic country-rocks, forming what are known as *migmatites*. The granite usually occurs as irregular veins of quartz and feldspar, separating thin stringers of highly metamorphosed country-rock. Often these veins are folded along with their country-rocks, while other veins may cut across the whole complex in an irregular fashion. The country-rocks are alway highly metamorphosed, so that they are now gneisses rather than schists.

The other granites found in the Scottish Highlands were mostly intruded after the peak of regional metamorphism had passed. Typically, they form discrete intrusions, often of large size, cutting across the structures found in their country-rocks. They vary widely in age, some dating back around 650 million years, while others were only intruded by the start of Devonian times, around 410 million years ago. This phase of igneous activity therefore lasted for over 250 million years. The intrusion of the most recent granites at least was accompanied by volcanic eruptions at the surface.

Old Red Sandstone

Resting unconformably on the Highland Schists are the Devonian rocks of the *Old Red Sandstone*. These sediments were deposited under continental conditions on the eroded roots of the Caledonian mountains, as they underwent uplift and erosion. The oldest rocks are found in the Southwest Highlands, where they were laid down as conglomerates and sandstones just before the eruption of the *Lorne Lavas* in Lower Old Red Sandstone times. These lavas are mostly andesites and basalts, up to 2,000 feet thick. Volcanic rocks of a similar age are found at Glencoe and Ben Nevis, preserved by cauldron-subsidence.

Depositional Environment. The Old Red Sandstone is also found in the Northern Highlands, along the western shores of the Moray Firth, and extending north into Caithness and the Orkney Islands. However, these are younger rocks, mostly belonging to the Middle Devonian, with only a thin sequence of Lower Old Red Sandstone conglomerates and breccias present at their base. They accumulated to a very great thickness in the sedimentary basin known as *Lake Orcadie*. This formed a vast sheet of shallow and often rather brackish water, to which the sea never had access.

Apart from breccias and conglomerates, which were deposited as screes and alluvial fans around its margins, the sequence consists mostly of flagstones. They were laid down as thinly-bedded repetitions of fine-

grained sandstone, siltstone and mudstone, as well as calcareous mudstone and flaggy limestone. The more calcareous horizons contain the
sparse remains of a great many primitive fish. Then, the whole sequence
was overlain by more massive sandstones, deposited by rivers bringing an
influx of coarser sediment into the area.

Triassic and Jurassic Rocks

Following the deposition of the Old Red Sandstone in Devonian times,
which ended around 360 million years ago, there is a long gap in the geological record of the Scottish Highlands. Rocks of Carboniferous age are
virtually lacking, while any Permian rocks are difficult to distinguish
from the overlying strata of Triassic age. Together, these Permo-Triassic
rocks represent the *New Red Sandstone*, found only locally along the
shores of the Moray Firth, but occurring in greater force around the
Tertiary Igneous Complexes of Skye, Mull and Ardnamurchan. Conglomerates and sandstones are found, passing up into Jurassic rocks.

Depositional Environment. The start of Jurassic times around 200
million years ago was marked by the sea spreading around the fringes of
the Scottish Highlands. A thick sequence of marine and some deltaic sediments was then laid down, particularly on Skye, consisting mostly of
sandstones, limestones and shales. The sandstones were deposited as
deltas, built out across the sea-floor, where otherwise limestones and
shales were accumulating. Later, estuarine or lagoonal conditions were
established by the middle of Jurassic times. All these rocks are often
packed with fossils, including corals, bivalves, ammonites, belemnites
and brachiopods.

A unique feature is seen around Portgower on the western shores of
the Moray Firth, where *boulder-beds* are found as an integral part of the
Jurassic sequence. They consist of a jumbled mass of angular boulders,
all belonging to the Middle Old Red Sandstone, which slid down a submarine fault-scarp as the result of earth-quakes.

Tertiary Igneous Activity

There are very few Cretaceous rocks preserved in the Scottish Highlands,
although it is thought that the Chalk was once deposited over much of the
area. Instead, *Tertiary Lavas* rest directly on the Jurassic sediments just
described from the Inner Hebrides. These rocks are witness to a dramatic
episode in the geological history of the Scottish Highlands, when volcanoes
erupted on Skye, Rhum, Ardnamurchan, Mull and Arran, around 60 million
years ago. All this igneous activity was closely associated with the initial
stages in the opening of the North Atlantic Ocean, as Northwest Europe part-
ed company with Greenland.

Tertiary Lavas. The earliest volcanic rocks are mostly basalt lavas and occasional tuffs, poured out over an ancient land-surface, so that they now lie unconformably on top of the older Jurassic rocks. They were probably erupted from volcanic vents, situated where the so-called *central complexes* are now found. These eruptions built up great thicknesses of lava-flows, exposed throughout much of Mull, Morvern and Northern Skye, together with the islands of Eigg, Muck and Canna, where they have been preserved from the effects of later erosion. Each lava-flow is rarely more than 50 feet thick, and one example has been traced for a distance of 13 miles. The tops of the lava-flows are often reddened by subaerial weathering under the tropical climate of Early Tertiary times.

Intrusive Complexes. This volcanic phase was followed by intrusion of the igneous rocks that now form the *central complexes* of Skye, Rhum, Ardnamurchan and Mull, as well as St Kilda and Arran. It is likely that this phase of igneous activity was accompanied by further outpourings of lava at the surface, but of this there is no direct evidence. Certainly, these intrusive complexes must represent the roots of ancient volcanoes as they continued to erupt.

Broadly speaking, they show an early phase of explosive activity, followed by the intrusion of basic and ultrabasic rocks like gabbro and peridotite. The closing stages in their evolution were often marked by the intrusion of more acid rocks such as granite. Dykes were also intruded in a northwesterly direction during this phase of intrusive activity, forming a *Tertiary dyke-swarm* that can be traced from Northeast England to the Outer Hebrides.

Pleistocene Glaciation

Since the ending of volcanic activity in Early Tertiary times, some 50 million years ago, the Scottish Highlands have been affected by widespread uplift and erosion. Since the sedimentary basin now formed by the North Sea subsided at the same time, it seems likely that these movements were accompanied by the whole land-mass tilting towards the east. Certainly, a land-surface was established in Late Tertiary times, sloping gently towards the North Sea, across which rivers flowed in the same direction from a watershed in the west. Renewed pulses of uplift probably resulted in quite a mountainous topography, which was then accentuated by the effects of the *Pleistocene Glaciation*. This began around 1.8 million years ago, and continued intermittently until the very end of what is known popularly as the Great Ice Age, just 10,000 years ago.

Glacial Erosion. The effects of this glaciation are seen nearly everywhere in the landscape of the Scottish Highlands. Glaciers in flowing downhill under the influence of gravity are able to attack the underlying rocks in two distinct ways. Firstly, the ice has a 'plucking' action, detaching blocks of rock along joints and other fractures, and then incorporating these blocks into the glacier as another addition to its load of rock fragments and other detritus. Secondly, the ice has a 'scouring' effect as it drags its load of rock fragments over the solid rocks forming its floor, grinding down this surface by abrasion.

Landforms produced by glacial erosion are a characteristic feature of the higher ground in the Scottish Highlands. Many valleys have the U-shaped profiles typical of *glacial troughs*, with steep hillsides sweeping down to form flat-bottomed valleys. These valleys are often eroded along their lengths into a series of *rock basins*, now occupied by lochs, often very deep, separated from one another by a series of rocky steps. Where these glacial troughs reach the coast, particularly in the west, they are likely to be occupied by arms of the sea, forming *fjords* or *sea-lochs*.

Although the glaciers eroding these valleys mostly flowed away from the higher ground in response to gravity, this was not always the case. In particular, the ice was at its thickest just to the east of the present-day watershed, forcing it to flow uphill to the west in some places. The glacial troughs eroded by this ice are then said to breach the watershed.

The heads of glacial valleys are often bounded by steep slopes, while *corries* have much the same shape, formed wherever the ice accumulated on the upper slopes of the mountains. Typically, they look like giant amphitheatres, backed by steep cliffs. Often, their floors are occupied by a small lake, dammed by solid rock or by glacial deposits. Where two corries occur close together, they may only be separated from one another by a narrow ridge known as an *arete*. Sharp peaks are formed wherever these ridges converge on one another, surrounded by corries which have cut into the mountain from several sides at once.

The effects of glacial erosion may also be seen wherever solid rocks are exposed. Often, these rocks are scoured and eroded by the ice to form *glacial pavements*. There are *glacial striae* sometimes visible on these surfaces, where slight grooves and scratches have been made by rock fragments as they were dragged along by the ice. Occasionally, these surfaces were moulded by the ice into more extreme forms, known as *roche moutonnées*. These have elongate and streamlined shapes, gracefully rounded with smooth contours facing upstream against the flow of the ice. Downstream, they often end steeply in an abrupt step, formed where the rock has been plucked away by the ice.

Glacial Deposition. While glacial erosion was the dominant process to affect the higher ground during the Last Glaciation, its place was taken by deposition once the ice flowed out beyond the mountains. **Boulder clay** was deposited widely over the lower ground, not only obscuring the pre-existing topography but also giving its own characteristic landforms. These deposits typically consist of scattered boulders of all shapes and sizes, lying in a gritty clay. It mostly accumulated below the ice-sheet, plastering the underlying surface with detritus.

Boulder clay often forms a dull and featureless landscape, only enlivened where it becomes more rolling in character, with slight hills and shallow valleys. Lochs are formed where the water cannot drain away from any slight depressions in such an undulating surface. Elsewhere, *drumlins* are found where boulder clay is thrown into smooth but elongated mounds, up to 100 feet high in some cases, giving rise to a landscape thought to resemble a 'basket of eggs.'

Much detritus was also dumped by the glaciers where the ice finally melted along its front. Where this ice-front remained stationary, so that the glaciers were neither advancing nor retreating, *terminal moraines* are found as distinct ridges of sand and gravel, often with larger boulders. More common are the much wider spreads of *hummocky moraine*, dumped where it lies as the glaciers gradually retreated, or just left by the melting away of "dead ice", once the glaciers had ceased to advance. As the glaciers retreated towards the end of the Great Ice Age, vast volumes of melt-water were released, carrying along sand and gravel as well as much finer material in suspension. The coarser detritus was deposited as *fluvio-glacial sands and gravels*, which often choke the lower reaches of the valleys draining the Scottish Highlands.

The Last Glaciation was also accompanied by the removal of sea-water from the oceans to form vast ice-sheets, which caused the Earth's surface to subside under their weight. Quite large changes in sea-level occurred as a result. However, the melting of the ice-sheets restored the sea to its original volume, while the land is still now only just recovering from the weight of the ice.

This means that the land is still rising, so that the sea once stood much higher in comparison with the land, just after the end of the Last Glaciation. *Raised beaches* are therefore a common feature of the coasts around the Scottish Highlands, particularly in the southwest. These features are best seen where they are cut into solid rock, forming wavecut platforms lying well above the present level of the sea, backed by ancient cliffs. Caves, natural arches and sea-stacks may all be preserved, together with beach deposits.

CHAPTER 4

Inverness to John O'Groats

THE TRAIL STARTS AT INVERNESS, following the A9 road for most of the way to John O'Groats, apart from the occasional detour. Although the scenery lacks the splendour seen in the Northwest Highlands, this part of the trail provides an excellent introduction to geology in the field. The route keeps to the low ground along the western shores of the Moray Firth, except where it climbs over the Ord of Caithness at the Sutherland border.

Nearly all this ground is underlain by Devonian rocks of the Middle Old Red Sandstone, except for a narrow outcrop of Triassic and Jurassic rocks between Golspie and Helmsdale. The Moine Schists make up the higher ground of the Northern Highlands to the northwest, apart from the Lower Old Red Sandstone, which locally forms the foothills to the higher ground. The Moine Schists are intruded by various granites, among which the Helmsdale Granite is the most conspicuous.

Much of the lower ground along the route is blanketed by boulder clay, giving rise to a typical landscape of smooth contours and very subdued features, which can be seen in crossing the Black Isle and the interior of Caithness, away from the coast. However, there are also widespread deposits of sand and gravel, laid down as the glaciers retreated, perhaps best seen in Easter Ross and farther north beyond the Dornoch Firth. Raised beaches are a common feature along the coast-line, occurring at various heights above sea-level. The lowest and most recent often gives rise to very wide areas of flat-lying land.

INVERNESS TO TARBAT NESS

Leave Inverness northbound on the A9, crossing the Beauly Firth by the Kessock bridge. The coast of the Black Isle to the northeast is remarkably straight, determined as it is by the line of the Great Glen Fault, lying just off-shore to the southeast. The effect of the fault-movements can be seen in the shattered nature of the Middle Old Red Sandstone conglomerates, where they exposed in roadcuts along the A9 just north of the bridge. Similar rocks are better exposed on the shore at *Craigton* [NH 663485], which can be reached by turning south off the A9 and driving east along the B9161 through the village of North Kessock itself.

Continue across the Black Isle, following the A9 to reach the road junction for Evanton on the north side of the Cromarty Firth, where a visit can be made to the **Black Rock Gorge**. This spectacular feature is reached by driving through the village of Evanton on the old A9, before turning left just beyond the bridge over the River Glass to follow the road which runs up Glen Glass. After about a mile, a rough track on the left leads down towards the river, which runs in a deep and very narrow cleft for a distance of nearly a mile.

A foot-bridge crosses the gorge at [NH 593668], and provides an excellent viewpoint. This gorge was cut by melt-waters towards the end of the Last Glaciation. Its fern-covered walls are formed by Old Red Sandstone breccias, crudely bedded and nearly horizontal. Return to Evanton, and turn left along the old A9 to regain the main road near Alness.

Alternatively, a visit may be made to **Hugh Miller's Birthplace** at Cromarty on the Black Isle, by taking the B9163 to the right just before the bridge over the Cromarty Firth is reached. Hugh Miller (1802-1856) was a stonemason to trade, with an world-wide reputation as a geologist. He eventually became the editor of an evangelical newspaper published in Edinburgh, which carried his essays on geology as well as book reviews and political articles. The cottage where he was born is now owned by the National Trust for Scotland. It houses a small selection of the fossils collected by him, as well as other relics of his remarkable life. A ferry runs in the summer across the Cromarty Firth to Nigg from where a cross-country route can be followed to reach the B9165 beyond Hill of Fearn.

On rejoining the A9 near Alness after visiting the Black Rock Gorge, continue north as far as its junction with the B9165, a few miles beyond Kildary. Turn right and then follow the B9165 through Hill of Fearn towards Portmahomack. Tarbat Ness is reached by taking the right fork just before entering this village. Stop in the car-park on the right of the private road leading to the lighthouse.

GEOLOGICAL LOCALITY: TARBAT NESS

Descend to a narrow inlet [NH 945873] with a storm beach at its back immediately northeast of the car-park. There is a rough path leading down to the shore. The rocks belong to the Middle Old Red Sandstone. They are mostly red sandstones with scattered pebbles of white quartzite, although occasional horizons are creamy yellow. The quartz grains forming these sandstones can easily be seen on a weathered surface using a hand-lens. The bedding is inclined away from the horizontal so that it now dips at 20° towards the northwest. Cross-bedding is well-developed, showing the sandstones were deposited by currents flowing from the southwest. Convolute bedding is also present, affecting the cross-bedded units.

Return to the car-park, and walk along the track past the lighthouse to the headland of Tarbat Ness itself. This provides a splendid panorama on a fine day, stretching from Buchan right round the shores of the Moray Firth to Caithness. The rocks exposed at the headland are pebbly yellow sandstones, quarried in the past for millstones, to judge by a reject lying close to the end of the track.

TARBAT NESS TO PORTGOWER

Returning through Portmahomack, take the right fork about 2 miles beyond the village, leading towards Tain. After passing through Tain, turn right along the A9, just beyond the town. Follow the A9 across the Dornoch

Firth, and continue north to cross the causeway at the head of *Loch Fleet*. It provides an excellent view of a glaciated landscape, where erosion has prevailed over deposition.

Strath Fleet has the typical U-shaped form of a glacial trough, even although its flat floor is covered by river deposits. The hills guarding its mouth are formed by rather massive breccias of the Middle Old Red Sandstone, scraped bare by the ice, leaving steep outcrops of glacially-scoured rock. Several of these hills have the form of 'roche moutonnées' on a grand scale, since ice flowing from the northwest has plucked away at the rocks to leave the steepest slopes facing southeast.

Stop just north of *The Mound* at a layby on the A9, about 200 yards beyond its junction with the A839 from Lairg, to examine the breccias forming these glaciated slopes where they are exposed at the roadside. They consist of angular fragments of impure grey quartzite and pink granite, set in a much finer-grained matrix of reddish purple sandstone. These pebbles have surfaces stained purple, and they are packed tightly together in the rock. These breccias were probably laid down as scree deposits along the flanks of a mountainous area, formed by the Highland Schists.

Drive north along the A9 through Golspie, where a stop can be made at the *Orcadian Stone Company*, situated on the main street just beyond the Bank of Scotland. There is a splendid exhibition of mineral specimens from all over the world, and a display of geological specimens illustrating the local geology. Books and maps can be purchased, as well as mineral and rock specimens.

The route now passes from the Old Red Sandstone on to a narrow strip of Triassic and Jurassic rocks, which lies along the coast for nearly 20 miles as far north as the Ord of Caithness. These rocks are mostly sandstones, limestones and shales. Coal was first mined from the Jurassic rocks at Brora in 1529, although this has now been abandoned. All these rocks are downthrown by the Helmsdale Fault against the more resistant rocks, forming the higher ground to the northwest. These older rocks mostly consist of the Middle Old Red Sandstone, except that the Helmsdale Granite appears farther northeast around Helmsdale. Much of the lower ground is mantled by thick deposits of sand and gravel, while there are long stretches of raised beach along the coast, particularly conspicuous at Brora..

Continuing north along the A9 to reach the village of *Portgower*, some 9 miles beyond Brora. Take the second turning to the right after entering the village. Then turn to the right after 50 yards, and drive to the end of the road, where cars may be parked.

GEOLOGICAL LOCALITY: PORTGOWER

It is essential to visit this locality at low tide. Walk 100 yards southeast down a track towards the sea from the end of the road, and follow this track as it turns right in its descent to the coast. After crossing the railway line with due care, walk southwest along the shore for 200 yards to reach the ruins of a wall. Just beyond this point, the coast forms a low promontory [ND 004127] running out to sea at low tide. Its farthest point is the Fallen Stack of Portgower.

The rocks are Jurassic in age. They consist of a series of boulder beds in which large blocks of Middle Old Red Sandstone rocks are embedded, surrounded by a matrix of shelly calcareous grit. These boulders are rather angular in shape, and vary greatly in size. They consist of rocks typical of the Caithness Flagstones, and carry fish remains as fossils which identify them as Middle Old Red Sandstone. An excellent example of just such a boulder bed is found immediately below high water mark, while another lies slightly farther offshore. This horizon carries large boulders up to 10 yards in length. Since the boulders consist of well-bedded rocks, it can be seen how they are completely disorientated, forming a higgledy-piggledy jumble of angular blocks. By way of contrast, the sequence as a whole dips southeast at a low angle.

By scrambling across these rocks, the Fallen Stack of Portgower can itself be reached. This forms a huge block of Middle Old Red Sandstone, measuring nearly 50 yards in length and 30 yards in width. Its bedding is close to the vertical, striking at a high angle to the overall dip of the beds. The interpretation placed on all these boulder beds is that they were deposited at the foot of a submarine escarpment in Jurassic times, which was first formed and then maintained by repeated movements along the Helmsdale Fault. Large blocks of Caithness Flagstone became detached at the top of this submarine escarpment, sliding and perhaps tumbling down this unstable slope into the deeper waters at its foot. This probably occurred in response to earthquake shocks along the Helmsdale Fault itself.

PORTGOWER TO JOHN O'GROATS

The route north from Portgower first passes through Helmsdale, where a detour can be made up the Strath of Kildonan to visit *Baile an Or* (the town of gold), where the gold-rush of 1869 took place. A permit to pan gold may be obtained from the estate office at Kildonan Farm, while there is a display of its history at the Timespan Heritage Centre in Helmsdale. Continue north along the A9 road from Helmsdale over the Ord of Caithness. Just beyond Navidale, typical exposures of Helmsdale Granite are seen on the hillside above the road.

Once the higher ground is reached beyond the old county boundary, the road crosses and recrosses the unconformity separating the Helmsdale Granite from the overlying Old Red Sandstone. Although this unconformity is exposed in a series of roadcuts at Ousdale, it is difficult to locate, as the overlying sediments are arkoses, looking much like the underlying granite from which they were derived. These sediments pass upwards into well-bedded sandstones and shales, which dip north at a gentle angle away from the underlying granite.

Once past Ousdale, the quartzite ridge of Scaraben comes into view with its grey screes, quite unlike the more rounded hills formed by the Helmsdale Granite to the south. The road then descends abruptly at Berriedale, only to climb back north to its previous level. Beyond are rolling moorlands of boulder clay, reaching the sea in steep if not vertical cliffs along the coast. Indeed, the coast is mostly inaccessible to the north of the Ord, except where rivers have cut deep valleys right down to sea-level. This landscape is typical of Caithness as a whole.

Nearly all the county is underlain by the Caithness Flagstones, belonging to the Middle Old Red Sandstone. However, these rocks are mostly mantled by thick deposits of boulder clay, except on the higher ground, so that inland there are few features of geological interest. Driving north, the Caithness Flagstones are well-exposed in roadcuts at the southern end of the **Dunbeath bypass**, just beyond the road junction [ND 154292]. The flat-lying bedding has a very flaggy appearance, while it is cut by sets of vertical joints, making the rocks appear like blocks of masonry.

Beyond Dunbeath, stop at the **Laidhay Croft Museum**, where a splendid panorama looks out towards the west. The rounded ridge of Scaraben is seen end-on, flanked to the north by the conical peaks of Maiden Pap and Morven. Scaraben is quartzite, belonging to the Moine rocks of the Scottish Highlands, while Maiden Pap and Morven are built of Old Red Sandstone conglomerates and sandstones, so that differences in geology are expressed very directly in the landscape.

Continue north along the A9 road past Thrumster, where a detour can be made to the coast at **Sarclet Haven** [ND 351433]. The rocks on the south side of the harbour are affected by a good example of a thrust fault. On entering the outskirts of Wick, turn right and follow the route signposted to the Old Castle of Wick. Park at the end of the road where it runs along the coast, just beyond the swimming pool, near the **South Head of Wick**.

GEOLOGICAL LOCALITY: SOUTH HEAD OF WICK

Walk down the slope formed by a bedding-plane to [ND 375493], where a vertical face is exposed just above high-water mark. Apart from the spectacular nature of this bedding-plane (MFG 4), dipping at a fairly shallow angle towards the north, the rocks also show a variety of other features. The vertical face already mentioned exposes thinly-bedded shales with some calcareous layers, passing upwards into grey siltstones and shales. Climbing up a near-vertical step reveals impressive arrays of sedimentary dykelets (MFG 63), which cut the shaly layers at a high angle to the bedding. They are composed of siltstone, penetrating downwards from thin beds of the same material.

These dykelets occupy what are known as synaeresis cracks, which are thought to form underwater by the contraction of muddy sediment in response to changes in salinity. Once formed, these gashes became filled with silty sediment, washed in when the overlying bed was deposited. Subsequently, these dykelets were folded as a response to the compaction which affected the whole sequence after its deposition. The dykelets (MFG 62) can be seen in plan where they are exposed on nearby bedding-planes. Although somewhat irregular, they tend to be roughly parallel to one another. They differ in this respect from sun-cracks, which typically have polygonal outlines, and which also occur on a much larger scale. Ripple-marks and occasional sun-cracks can also be seen on these bedding-planes.

The rocks are also well-jointed with two sets of vertical fractures cutting across the bedding, trending east-west and northwest-southeast. Thin veins of light-coloured calcite and brown-weathering dolomite occupy the latter set of veins. Opposite a small inlet, just a short distance to the north of the exposures just described, there is a vertical fault in the cliff, marked by the presence of veins of calcite and dolomite. The downthrow on this fault is 3 feet to the south, to judge by its effect on a massive bed at the top of the cliff.

Return to the main road, and take the A9 through the centre of Wick towards John O'Groats. The country north of Wick is low-lying, and sand-dunes are well-developed along the shores of Sinclair Bay. Farther north, however, the road climbs around Warth Hill, where the Orkney Islands are seen across the waters of the Pentland Firth. On reaching John O'Groats, park near the Last House.

GEOLOGICAL LOCALITY: NESS OF DUNCANSBY

To reach the Ness of Duncansby [ND 391738], walk east along the coast from the Last House. The John O'Groats Fish Bed is exposed on the shore below high-water mark around 120 yards from the pier. It forms a calcareous horizon, creamy brown in colour. The rocks are otherwise mostly dull red sand-stones, dipping northeast at a low angle, interbedded with reddish shales and siltstones. These rocks lack the flaggy character which is typical of the Caithness Flagstones.

Passing around a slight bay, backed by a beach of shell sand, igneous rocks are first encountered just below high-water mark, some 50 yards beyond an old windlass. They are poorly exposed, forming a dyke of very dark rock, almost black in colour, which appears to cut across the bedding of the sandstones. Continue around the headland itself, beyond which a vol-canic vent is clearly seen on the foreshore, surrounded by red sandstones.

The exposures found just below high-water mark at the edge of the beach clearly reveal the nature of the agglomerate lying within this vent. It forms a breccia with angular fragments up to 6 inches in length, set in a much finer-grained matrix of comminuted material, grey or purple in colour. Fragments of dark igneous rock are conspicuous, together with large pieces of an almost black mineral, probably augite, consisting of single crystals with cleavage planes that glint in the sun. There are also fragments of sandstone, limestone and gneiss. Although the contacts of this vent are not well-exposed, it forms an area about 200 yards across, quite distinct from the sandstones which make up its country-rocks.

GEOLOGICAL LOCALITY: DUNCANSBY HEAD

The cliffs around Duncansby Head are vertical or even overhanging, and due care should be taken. Walk east around the coast to the Bay of Sannick, where the red sandstones are cut off by a fault. This strikes south of south-east across the peninsula separating Duncansby Head from the ground to the southwest. It brings up the Caithness Flagstones to the surface, where they form impressive cliffs around the headland itself.

Deep geos with vertical or even overhanging walls penetrate the headland where the sea has attacked the land along joints or faults in the flagstones. There are sea-caves at the back of these inlets wherever the sea is still active in its attack upon the land. Gaping chasms are found wherever the roofs of these caves have collapsed, leaving a bridge of rock across their seaward ends. As the cliffs continue to recede, sea-caves often join up to form natural arches, prop-ping up the cliffs like flying buttresses. Where the arch itself collapses, a sea-stack is left standing offshore as a testimony to the power of the sea in wearing away the land. All these features can be seen in walking around Duncansby Head from the Bay of Sannick to the small bay [ND 403728] lying to its south, beyond The Knee.

Descend to the shore at the back of this small bay by a steep and rather rough path. This leads down a gully from the iron gate in the fence that skirts the cliff-top at this point. This gully is eroded out along the fault which cuts across the peninsula of Duncansby Head. On reaching the shore, it can be seen that the rocks exposed by the cliffs to the left of this gully are rather massive red sandstones, similar to those found around John O'Groats. They are quite different from the typical flagstones that make up the cliffs on its other side. Since the bedding is flat-lying on both sides of the gully, a fault most likely runs down the gully, even though it is not exposed at this level.

In fact, just such a fault is exposed on the foreshore at low tide, where it is marked by a zone of shattering, about 10 feet in width. Looking north of northwest, it is clear that the flagstones turn down against this fault from the right, while the sandstones turn up as they approach the fault from the left. This suggests that the sandstones have been thrown down against the flagstones from a higher level, so that they are the younger rocks. The sandstones are also seen to be faulted where they are exposed in the cliff-face, 15 yards to the southwest of the gully. This fault dips steeply northeast, and downthrows about 8 feet in the same direction. This means that it is a normal fault, although it has the opposite sense of downthrow in comparison with the much larger fault that runs down the gully itself. Return to the car park at John O'Groats.

CHAPTER 5

John O'Groats to Cape Wrath

THE NORTH COAST OF CAITHNESS and Sutherland provides an excellent cross-section through the Northern Highlands. Starting from John O'Groats, the trail first crosses the Middle Old Red Sandstone rocks which we have already encountered farther south. These sediments rest unconformably on the metamorphic and igneous rocks of the Scottish Highlands, quite spectacularly in places like Red Point and Portskerra. The trail then crosses the outcrop of the Highland Schists, where they reach the wild and very rocky northern coast of Sutherland. These rocks are very complex. Although Moine Schists are present, they are associated with basement rocks of Lewisian aspect, which have undergone so much deformation and metamorphism that their original nature is open to doubt.

Farther west, however, the geology becomes less complex. There, the Moine Schists are found over a wide area, separated by fold-cores and thrust-slices of the Lewisian basement, now greatly reworked by the Caledonian deformation and metamorphism. All these rocks are thrust west of northwest over the underlying rocks of the Northwest Highlands by the Moine Thrust. This thrust outcrops around Loch Eriboll. Beyond lies a belt of disturbed rocks, carried forward of the thrust itself. The trail then reaches the northwestern foreland to the Caledonian belt, where Lewisian Gneiss, Torridonian Sandstone and the Cambro-Ordovician rocks are all preserved from the effects of the Caledonian earth-movements.

JOHN O'GROATS TO PORTSKERRA

Follow the A836 road west from John O'Groats towards Thurso. All this ground is covered by a thick blanket of boulder clay, which mostly obscures the underlying rocks of the Middle Old Red Sandstone except where they are exposed along the coast. At the village of Dunnet, a detour can be made to *Dunnet Head*, which is the most northerly point on the Scottish mainland. The rocks outcropping around this headland are red and yellow sandstones of the Upper Old Red Sandstone. They are much more resistant to weathering and erosion than the Middle Old Red Sandstone, against which they have been faulted. These sandstones can best be examined on the shore at the *Point of Ness* [ND 209711], where they display much cross-bedding, often affected by slumping. The cliffs north of this point clearly show the well-bedded character of these rocks.

Continue west towards Thurso around the shores of Dunnet Bay with its splendid fringe of sand-dunes, facing out over the waters of the Pentland Firth towards the northwest. Between Castletown and Thurso, the road runs inland, but a detour can be made to visit *Clairdon Head* around [ND 138700], where sun-cracks (MFG 61) are particularly well-developed. This locality is best approached by taking the side road passing through West Murkle to the Haven.

After passing through Thurso on the A836, a good view can be seen from near the caravan site [ND 110689], across the bay to Scrabster Harbour. The harbour itself is backed by cliffs of boulder clay, forming steep slopes cut by gullies. The great thickness of boulder clay can easily be appreciated, as virtually no solid rock can be seen, except at the very foot of these slopes.

Follow the A836 road past Dounreay to the village of Reay. The scenery changes once the sedimentary rocks of Caithness give way to the igneous and metamorphic rocks of the Scottish Highlands. This is marked around Reay by knolls of igneous rock, which appear as if stripped of their sedimentary veneer of Middle Old Red Sandstone rocks. Park at the viewpoint [ND 933646] on the north side of the road, just over 2 miles west of Reay.

GEOLOGICAL LOCALITY: RED POINT

Although it can only be reached by walking over rough moorland, this splendid locality [ND 930659] easily repays the effort involved. Start along a poor path 50 yards west of the layby, which heads north towards a heather-covered hillock in the distance. After nearly half a mile, follow a stream which runs northwest down a slight valley towards the sea, aiming for a grassy knoll. On reaching the cliff-top, about quarter of a mile to the east of Red Point, turn right and walk northeast towards the headland. **Great care should be taken where walking along these cliffs, since they are vertical, capped with steep and slippery slopes of boulder clay.**

Well-bedded rocks of the Middle Old Red Sandstone are exposed along this coast, dipping gently towards the north. Offshore, there are several sea-stacks. Just before Red Point, the rocks can be examined in safety, once the cliffs fall away towards the sea. The exposures just east of Red Point show grey fine-grained limestones overlain by coarse breccias. All these rocks are underlain by the flaggy sediments found farther west along the coast.

Traced east towards Red Point, these flagstones are replaced by coarse breccias, as can be seen on reaching the headland itself. These breccias weather out as very knobbly surfaces. They consist of very angular fragments of granitic rock, up to a foot across at the very most, together with less common fragments of quartzose schist and vein quartz. These fragments are usually packed together in a felspar-rich matrix of gritty arkose. Locally, however, very fine-grained limestone acts as a cement.

All these rocks are banked unconformably against the coarse-grained granite which is exposed at Red Point itself. The actual contact is close to the vertical, and it can be traced right to the top of the cliffs. It is best exposed on a ledge running out to sea below a grassy slope, some 50 yards to the west of the headland. A thin selvedge of breccia can be traced along the contact, where it is plastered against the granite. Tongues of breccia can be traced along the bedding for short distances away from this contact, before wedging out as the breccias are replaced by flaggy sediments.

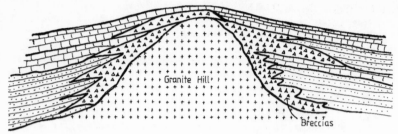

Figure 5. Vertical Cross-Section at Red Point.

Beyond this contact, granite is exposed as far as the headland itself, together with some inclusions of gneissose schist, cut by granitic veins. Just east of the headland, breccias are again banked against the granite, wedging out as they are traced away from the unconformity. These fragmental rocks are interbedded with impure limestones, forming the bulk of the sequence at this point. However, these limestones are themselves underlain by quartzose sandstones carrying much detrital feldspar, evidently derived from the underlying granite.

Clearly, the sedimentary rocks at Red Point buried a steep-sided hill of much older granite in Middle Old Red Sandstone times, which has now been partly exhumed from beneath its cover of screes, sandstones and flaggy limestones. Return to the road by walking south of southeast until the path is reached at the head of the shallow valley.

To reach the next locality, drive west along the A836, crossing the Halladale River with its terraces of fluvioglacial sands and gravels. After passing through Melvich, turn right at the hotel to reach Portskerra. Fork right, and park at a small group of houses near the coast. Walk down the track leading to the harbour.

GEOLOGICAL LOCALITY: PORTSKERRA

This locality marks the western margin of the Orcadian Basin, where the Middle Old Red Sandstone can be seen resting unconformably on top of the metamorphic rocks of the Scottish Highlands. The irregular nature of this unconformity (MFG 108) can best be appreciated where it is exposed in the cliffs [NC 877665] to the north of the harbour, capped with boulder clay. The underlying rocks are schists and gneisses, intruded by granitic veins. They are exposed as two knolls, lying on either side of the bay. These rocks, pink in colour and lacking any bedding, are overlain unconformably by breccias and sandstones, belonging to the Middle Old Red Sandstone.

The well-marked bedding shown by these rocks is quite clearly draped over an irregular surface cut by erosion across the underlying rocks, forming a shallow syncline. Evidently, the sedimentary rocks buried an ancient landscape of low hills and valleys. By scrambling across the foreshore, this unconformity can be seen at close quarters (MFG 105). The metamorphic rocks are overlain by rubbly material, which passes up into a basal breccia. Granitic fragments are present in this breccia, clearly derived from the physical disintegration of the underlying rocks.

PORTSKERRA TO LOCH ERIBOLL

Returning to the main road, drive west along the A836 towards Bettyhill. Although the road first crosses ground heavily mantled with boulder clay, once beyond Strathy this gives way to a much more rocky landscape, where the glaciers have left little behind in the way of superficial deposits. Strathy itself lies on an outlier of Middle Old Red Sandstone, mostly sandstones and conglomerates, but the underlying metamorphic and igneous rocks appear in force farther to the west.

These basement rocks can be seen by taking a rough track to the top of *Cnoc Mhor* [NC 757638], which leaves the main road just before the turning for Kirtomy, some 2¹/₂ miles beyond Armadale. The exposures along the track just below the summit show excellent migmatites, in which irregular but very abundant veins of granitic material are intimately associated with coarse-grained but still schistose country-rocks (MFG 243, 244 & 246).

The view from Cnoc Mhor is spectacular on a fine day, stretching from Dunnet Head and the Orkney Islands in the east to Whiten Head and the mountains of Ben Loyal, Ben Hope and Foinaven in the west. The twin peaks of Ben Griam Mhor and Ben Griam Beg stand out in the south as outliers of Middle Old Red Sandstone conglomerate, while Morvern is seen on the distant horizon. The extraordinary nature of the low plateau which makes up most of North Sutherland can clearly be appreciated from this view-point: it probably marks a surface of erosion exhumed from beneath a sedimentary cover of Old Red Sandstone rocks.

Return to the A836, and continue west through Bettyhill, where a well-developed terrace can be seen, lying around 50 feet above sea-level at the mouth of the River Naver. This forms the seaward end of a series of river terraces, constructed from fluvio-glacial sands and gravels, which can be traced in an irregular fashion along Strathnaver. Crossing the higher ground to the west, a detour can be made through Skerray by turning right along a side road to the north just after the A836 crosses the River Borgie.

This road follows a river terrace which can be traced down the River Borgie to the sea. Another terrace can be seen at exactly the same height on the other side of the valley. This terrace opens out to form a wide area of flat-lying ground, overlooking the mouth of the river around 50 feet above sea-level. Its seaward side is flanked by sand-dunes. This is a characteristic feature of this very exposed coast, where strong winds blowing off the sea can carry sand inland to heights of 400 feet, encouraging interesting communities of unusual plants as a result.

Stop at a parking place [NC 680611], just after the road forks, and cross a footbridge over the River Borgie to reach *Torrisdale Beach*. The outcrops exposed to sand-blast at the back of this beach around [NC 689618] display impressive examples of folding (and boudinage), which affect a series of granitic veins cutting across the metamorphic rocks at this locality.

Return to the road, and continue through Skerray to reach the main road, two miles east of Coldbackie. A stop can next be made just after this small hamlet, parking at the layby above *Coldbackie Bay* [NC 610601]. The roadcut just opposite this layby exposes excellent examples of fold mullions (MFG 226) in quartzo-felspathic schists, typical of the Moine rocks in this area.

There is also a dramatic view of the steep cliffs formed by the massive conglomerates of Watch Hill, or Cnoc an Fhreiceadain. Although once thought to be Devonian, a Permo-Triassic age now appears more likely for these conglomerates. They can be examined in a shallow gully, just to the left of the path as it descends towards the beach. The unconformity made by these conglomerates with the underlying Moine Schists can be located within a few feet in a series of exposures at the back of this beach, close to its eastern end.

Continuing west, the main road becomes the A838 just before the village of Tongue. This road then crosses the Kyle of Tongue by a causeway. Ben Loyal makes a very conspicuous feature to the south with its deep corries and sharp rocky peaks. Its summit ridge is crowned by tors, making it unusual among Scottish mountains. It is formed by an intrusive mass of syenite, an igneous rock much like a true granite but lacking any quartz. Its distinctive features reflect its geological structure, quite unlike that of its surrounding country-rocks.

These are the Moine Schists, which have taken on an easterly dip as the Moine Thrust is approached from this direction. This is also reflected in the surrounding landscape, which consists of a series of steep scarps facing towards the west, backed by gentler dip-slopes descending towards the east. Ben Hope is merely the highest of these scarps, somewhat disguised by the deep corries on its eastern flanks.

West of the causeway, a detour can be made around the Kyle of Tongue to examine garnetiferous mica-schists in the exposures around the broch at **Kinloch**. Take the side road south along the western side of the Kyle of Tongue, and park at a sharp bend in the road [NC 554531], just below the broch where it overlooks the Kinloch River. Typical garnet-mica-schists are well-exposed along the path leading uphill to the broch. The deep-red garnets resist weathering to stand out proud of the surface. Continue along this road through Kinloch to rejoin the main road at Tongue.

Another detour can be made along the western shore of the Kyle of Tongue by taking the side-road north through Melness and Talmine. The bedding of the Moine Schists is folded where these rocks are exposed in the crags overlooking the road just east of **Loch Vasgo** [NC 581645], while the quartz-rodding (MFG 227) which affects the Moine rocks exposed around the summit of **Ben Hutig** [NC 538653] can be seen by climbing this hill from Achininver.

Back at the main road, drive west along the A838 across the wide and rather featureless moorland of A'Mhoine. Although the Moine Schists are poorly exposed, they are named after this locality, where they were first mapped by the Geological Survey. Their outcrop is bounded to the northwest by the Moine Thrust, which is crossed as the road descends towards Loch Hope. Although not exposed, it is known to dip towards the east-southeast at a shallow angle.

The landscape changes dramatically to expose much more rock as this important dislocation is crossed. The first exposures encountered are Lewisian Gneiss, which forms the crags above Loch Hope. These rocks have been carried west on the Arnaboll Thrust, which lies underneath the Moine Thrust itself. The rocks lying below below this thrust, and outcropping to its northwest, are Cambrian quartzites belonging to the Pipe Rock. After crossing the bridge over the River Hope, these quartzites may be examined north of the A838, where they are well-exposed on the slopes of **Ben Heilam.**

The Arnaboll Thrust itself is exposed on Ben Arnaboll, where it makes a prominent feature in the crags overlooking the main road, a mile west of the River Hope (see Figure 6). The road then reaches a wide outcrop of Durness Limestone which runs along the eastern shore of Loch Eriboll. This limestone is first seen at a bend in the main road just after it passes Loch Ach'an Lochaidh. As **Loch Eriboll** comes into view, stop at the layby [NC 453601] overlooking Ard Neackie.

GEOLOGICAL PANORAMA: LOCH ERIBOLL

Looking south towards the head of Loch Eriboll from this viewpoint provides an excellent cross-section across the Moine Thrust-Belt. It shows how the Caledonian earth-movements have reversed the original succession of the rocks, bringing Cambrian Quartzite and Pipe Rock on top of Durness Limestone, and then Lewisian Gneiss on top of Cambrian Quartzite and Pipe Rock.

The hills lying beyond Loch Eriboll are Cambrian Quartzite, resting unconformably on top of Lewisian Gneiss. These rocks mark the foreland to the Moine Thrust, unaffected by the Caledonian earth-movements. The eastern slopes of Carnstackie, Beinn Spionnaidh, Meall nan Cra and Meall Meadhonach are all dip-slopes of Cambrian Quartzite, capped in some cases by Pipe Rock, descending towards the shores of Loch Eriboll. Only the much rougher outlines of Beinn Ceannabeinne, lying just out of sight to the northeast, reveal the underlying Lewisian Gneiss.

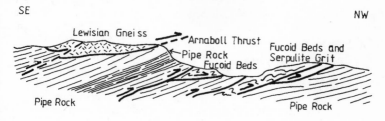

Figure 6. Vertical Cross-Section through the Ben Arnaboll Thrust.

The lowermost thrust to affect the rocks of the Moine Thrust Belt lies hidden under the waters of Loch Eriboll. It brings Durness Limestone on top of Cambrian Quartzite and Pipe Rock. Where it is overlain by boulder clay around Eriboll House, this limestone gives fertile ground capable of cultivation. Its outcrop covers much of the lower ground along the eastern shores of Loch Eriboll, running out to sea at An t'Sron. However, this formation also makes up Ard Neackie, where limekilns reveal its presence, while it continues northwards as a thin strip along the coast towards the lighthouse at the mouth of Loch Eriboll. It can be examined on the slopes of Torr na Bithe, just to the north of the viewpoint itself, where it weathers out as a typical limestone.

The Durness Limestone gives way to older rocks of the Cambrian sequence towards the east. Although the Pipe Rock is well-represented, the Fucoid Beds and Serpulite Grit are also present in some places, together with the Basal Quartzite. All these rocks are affected by folding and thrusting to such an extent that they are locally turned upside down. These structures are best seen around Ben Heilam, north of the main road. Traced to the south, all these rocks become hidden under the Lewisian Gneiss, where this formation is carried forward on the Arnaboll Thrust. This thrust reaches the coast just to the south of the viewpoint, where it carries on its back not just Lewisian Gneiss but also Basal Quartzite and Pipe Rock. The Basal Quartzite forms the conspicuous craigs overlooking the road just south of the viewpoint, while the Pipe Rock is exposed along the road at the back of Camas an Dun. All these rocks are slightly overturned towards the northwest.

The Lewisian Gneiss lying on top of the Arnaboll Thrust overlooks much of the lower ground towards the head of Loch Eriboll. It makes the far crags of Beinn Arnaboll on the skyline to the southeast, lying beyond the exposures of Cambrian Quartzite just mentioned, while it also forms the prominent crags of Creag na Faoilinn, overlooking the head of Loch Eriboll. The original character of these gneisses has been destroyed in many places by the intensity of the earth-movements, converting them into the flaggy and very fine-grained rocks known as mylonites.

All these rocks are capped by the Moine Thrust, which runs just below the skyline, east of Loch Eriboll. It does not make a very conspicuous feature in the landscape, although it brings forward the Moine Schists on top of the underlying rocks. The shallow dip to the southeast of the rocks lying above the Moine Thrust is clearly reflected in the slope of the skyline forming the summit of An Lean-charn in the far distance.

LOCH ERIBOLL TO CAPE WRATH

Continue west towards Durness along the A838 around the head of Loch Eriboll, where Creag na Faoilinn gives a much closer view of the Lewisian Gneiss, thrust over the underlying rocks. There is a wide terrace of fluvioglacial gravels at the mouth of Strath Beag. Its surface is pock-marked by several lochans, which occupy 'kettle-holes' where masses of 'dead ice' have been left to melt away as the glaciers retreated at the end of the Last Glaciation. A stop can be made opposite *Polla* to take in the geological panorama up Strath Beag, where the effects of the Caledonian earth-movements are particularly clear.

Beyond this point, the road runs along the foot of the dip-slope formed by the Cambrian Quartzite where it is exposed to the northwest of Loch Eriboll. Beyond the turning to Portnancon, the road crosses a faulted contact made by these quartzites with the underlying gneisses, which form the higher ground around *Beinn Ceannabeinne*. The effects of glacial action are clearly seen 300 yards west of the road around [NC 440642], where there is a good example of a glacial pavement, and north of the side road leading to *Rispond*, where superb examples of roches moutonnées can be seen around [NC 451656].

Just beyond the turning to Rispond, stop at the car-park overlooking the beach at *Ceannabeinne* [NC 443654]. The nearby exposures show banded gneisses (MFG 254), which are typical of the Lewisian Gneiss where it has

46

been affected by the Laxfordian movements. Other good exposures are found at the western end of Ceannabeinne Beach [NC 441657], where darker layers of basic rock are present in the gneisses, breaking up into segments as the result of boudinage.

Returning to the road, glacial striae are visible on a glacially-smoothed surface of Lewisian Gneiss which faces the road where it bends sharply to the right, some 200 yards west of the car-park. A perched block of Lewisian Gneiss, left behind by the ice when it melted away, makes a prominent feature after another 200 yards, just north of the road where it reaches the top of the hill.

The road then crosses a short stretch of country typical of the Lewisian Gneiss, before encountering Cambrian Quartzite and Pipe Rock around Sangobeg. The bedding shown by these quartzites can be seen in the low cliffs running out to sea at the next headland, just before Lerinmore is reached. These rocks are faulted against Durness Limestone, which is exposed at the roadside on approaching *Smoo Cave* [NC 419671].

The waters of the Allt Smoo now descend 70 feet into the Smoo Cave by disappearing down an impressive sink-hole, just south of the road. Over the years, this stream has dissolved away the limestone and dolomite to form the cave itself. It has much the largest entrance of any such cavern in Britain, 100 feet across and 50 feet high, and inside there is a second cave with an underground lake. Smoo Cave lies at the back of a long and narrow inlet of the sea, with very steep walls, forming a gorge-like feature. Most likely, there was once a roof to this gorge, which then collapsed as the sea continued to pentrate inland along the line of the Allt Smoo.

Driving west along the A838 towards Sango Bay, more exposures of Durness Limestone are seen at the roadside. *Sango Bay* is underlain by shattered quartzites, schists and gneisses, lying on top of the Durness Limestone. The presence of these rocks might be thought a mystery until it is realised that they are an advance guard of the Moine Schists, forming what is known as an outlier. They were thrust up the Moine Thrust high above the present level of erosion, and then dropped down by normal faulting at a much later date into their present position.

Another outlier of Moine Schists is seen at Faraid Head, some 10 miles across country from the outcrop of the Moine Thrust itself at Loch Eriboll. The movements on this thrust-plane must therefore have exceeded this figure. In fact, they were probably very much greater, as the rocks lying on top of the Moine Thrust have been thrust for many tens of miles across country from the east.

The normal faulting around Durness has preserved the Durness Limestone from the effects of erosion, well in advance of its main outcrop below the Moine Thrust at Loch Eriboll. One such fault is marked by the steep wall of Durness Limestone, which runs along the southeastern side of Sango Bay. It is this fault which brings down the metamorphic rocks to the northwest around the shores of this bay.

There is another normal fault running southwest from Sangobeg. It crosses the low ground lying at the foot of the hills between the Kyle of Durness and Loch Eriboll. This fault throws Durness Limestone down to the northwest against Lewisian Gneiss to the southeast. The gneiss makes the higher ground to the southeast, capped by Cambrian Quartzite.

After passing through Durness, take the side road which leads to the ferry across the Kyle of Durness. Although this is just a passenger ferry (operating only in the summer months), it links up with a fast and furious minibus service to Cape Wrath. The route passes across the Lewisian Gneiss for most of the way, apart from an outlier of Cambrian Quartzite and Torridonian Sandstone around Daill.

However, the hills north of the road are Torridonian Sandstone, faulted against Lewisian Gneiss to the southwest, and capped with Cambrian Quartzite on Sgribhis-Bheinn. These hills reach the north coast at Clo Mor, where the Torridonian Sandstone is exposed in the highest cliffs on the British mainland, dropping over 900 feet into the sea, and stretching for nearly 2 miles along the coast. The hills south of the road are likewise Torridonian Sandstone, which also forms most of the coast-line south of Cape Wrath.

GEOLOGICAL VIEWPOINT: CAPE WRATH

Although not so high as Clo Mor, the cliffs around Cape Wrath [NC 260747] rise to well over 400 feet, just east of the lighthouse. They provide spectacular views of the Lewisian Gneiss, showing that its banded nature is the result of alternations of light-coloured gneiss with much darker layers of basic rock, while the whole complex is cut by later sheets of granite and pegmatite. All these rocks are folded to a certain extent. There are good views out to sea on a good day, with the Butt of Lewis and North Rona visible on the far horizon.

CHAPTER 6

Cape Wrath to Ullapool

SOUTH OF CAPE WRATH LIE the Northwest Highlands of Scotland, which acted as a foreland to the Caledonian earth-movements, northwest of the Moine Thrust. This is classic ground, visited by generations of geologists, while even the casual tourist cannot fail to be impressed by how the geological foundations to the country are clothed in landscape and scenery. The Northwest Highlands are bounded on the east by the Moine Thrust, which carries the Moine Schists of the Caledonian Highlands northwest over this ancient foreland.

This great dislocation was accompanied by much folding and thrusting of the underlying rocks, along the southeastern edge of the Northwest Highlands. These rocks now form a belt of structural complication, several miles in width, lying in advance of the Moine Trust itself. Among the other thrusts recognised as important elements of this thrust belt are the Glencoul, Ben More and Kishorn Thrusts, together with the lowermost Sole Thrust, below which the rocks lying farther to the northwest are not affected in any way by the Caledonian earth-movements.

The Northwest Highlands consist of Lewisian Gneiss, Torridonian Sandstone and Cambro-Ordovician rocks, and it is these rocks which are encountered within the Moine Thrust Belt, shuffled together by the Caledonian earth-movements. Farther northwest, they are not affected by these movements, and their original character can best be appreciated. The Lewisian Gneiss is exposed over wide areas as a very rough terrain of low rocky knolls and peaty hollows, strewn with a great many lochans. The Torridonian Sandstone gives rise to a quite different scenery of high mountains with many peaks, often rising abruptly in terraced slopes from a foundation of Lewisian Gneiss, and ringed around with great precipices. These mountains of dark-red sandstone are often capped by glistening white Cambrian Quartzite and Pipe Rock, which adds another distinctive feature to the landscape.

CAPE WRATH TO LAXFORD BRAE

Returning from Cape Wrath across the Kyle of Durness, continue south along the A838 towards Scourie. The wide valley of the River Dionard is floored at first by the Durness Limestone, and typical exposures are seen where the road crosses the river. The hills to the southeast are Lewisian Gneiss, capped at their very top by Cambrian Quartzite. Although Lewisian Gneiss also occurs northwest of the road, this gives way to Torridonian Sandstone as the road starts to climb towards Gualin House. This sandstone with its massive bedding forms the slopes of Farrmheall to the north, capped by Cambrian Quartzite.

Once **Gualin House** is reached, it can be seen just how rugged a terrain is formed by Lewisian Gneiss, particularly where it makes the lower ground below Foinaven, southeast of the road. This makes a strong contrast with much more subdued topography to the northwest, which marks the outcrop of the Torridonian Sandstone with its lack of rocky exposure. The valley itself roughly follows the faulted contact separating these two geological formations from one another.

The shapely peak of Foinaven faces out northwest over this ground. Although its summit and eastern flanks are Cambrian Quartzite and Pipe Rock, its northernmost peak is Lewisian Gneiss. Its northwesterly slopes, like those of Carnstackie to the northeast and Arkle to the south, are all formed by Lewisian Gneiss.

As the road descends towards the head of Loch Inchard at Rhiconich, it enters extremely craggy ground typical of the Lewisian Gneiss throughout the Northwest Highlands. Stop in a layby at the locality known as Laxford Brae, 2 miles south of Rhiconich.

GEOLOGICAL LOCALITY: LAXFORD BRAE

The exposures along the road at this locality [NC 235901] provide an excellent section through the Lewisian Gneiss where it has been affected by the Laxfordian movements, showing the structural complexity typically displayed by these rocks. They are banded gneisses, intruded by much pegmatitic material, which forms the light-coloured masses of granitic rock, invading what must be older rocks in a very irregular fashion. The banded gneisses are themselves formed by alternations of different rock-types. The lighter layers are grey gneisses, approaching granite in composition if somewhat less acid, while the darker bands are much more basic rocks with the composition of basalt or gabbro.

Although the various layers in these banded gneisses often appear quite parallel to one another, careful examination shows that this is not always the case. In fact, a thin basic layer can be seen towards the northern end of the section, cutting across another basic layer at a slight angle (MFG 253). This means that two generations of basic rock are present within the Lewisian Gneiss at this locality. The cross-cutting layer may well represent the remnants of a Scourie Dyke, now almost incorporated into these Laxfordian Gneisses.

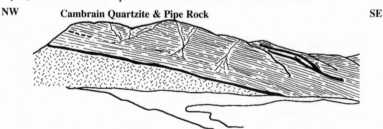

NW   **Cambrain Quartzite & Pipe Rock**   SE

Figure 7. Unconformity between Lewisian Gneiss and the Cambrian Rocks on Arkle.

LAXFORD BRAE TO LOCH GLENCOUL

Continue south towards Laxford Bridge, passing typical exposures of Lewisian Gneiss where the road reaches the shores of Loch Laxford. Turn left at Laxford Bridge along the A838 to make a detour to Loch Stack: otherwise turn right along the A894 towards Scourie. By parking at the southeastern end of Loch Stack around [NC 293416], a fine view can be obtained to the north, where the Cambrian Quartzite and Pipe Rock rests unconformably on top of Lewisian Gneiss along the southern flanks of Arkle. Like the bedding of the overlying sediments, this unconformity is inclined towards the southeast at a low angle. Traced down-dip in the same direction, it can be seen how these rocks get caught up by the Caledonian earth-movements, so that they become increasingly disturbed in this direction.

Return to Laxford Bridge and continue west along the A894 towards Scourie, crossing the geological boundary between the Scourian and Laxfordian elements within the Lewisian complex as a whole. It is marked by a wide zone of granitic sheets, which runs northwest-southeast along both shores of Loch Laxford. The much older Scourian Gneisses lie beyond this zone to the southwest, where they can be seen around Scourie.

Even the casual visitor can appreciate this change in the geology since the Scourian Gneisses lack much in the way of granite, unlike the Lewisian Gneisses of Laxfordian age to the northeast. This gives a rather gloomy feel to the landscape around Scourie and farther south, often marked by drab outcrops of rather dark rock, particularly in the deep roadcuts along the A894. Farther north, the landscape feels more cheerful as there is much more granite, forming light-coloured exposures glinting in the sun.

Three coastal sections can be recommended to the visitor who wants to get to grips with the geological complexities of the Lewisian Gneiss. The first lies north of *Tarbet* [NC 489163], where the Scourian Gneisses give way to the Laxfordian rocks across what is known as the Laxford Front, itself marked by the zone of granite sheets along Loch Laxford. The second lies south of Scourie, where a Scourie Dyke is exposed on the coast around [NC 145415] near *Upper Badcall*, cutting across Scourian Gneisses where they are little affected by the Laxfordian movements. Excellent exposures of garnetiferous pyroxene-granulites are seen on the coast 1½ miles north of the last locality at [NC 143442], best reached from *Scourie Mor*. All three localities are fully described in the Geologists' Association *Guide to the Lewisian and Torridonian Rocks of the North-West Highlands*, which can be obtained at local bookshops.

Driving south along the A894 from Scourie, Quinag comes into view, just before Kylesku. This mountain is nearly all Torridonian Sandstone, magnificently exposed in the great buttresses of Sail Gorm and Sail Garbh, which face out over the Lewisian Gneiss to the north. There is a capping of Cambrian Quartzite on its highest summit, while this quartzite also forms a great sheet which descends its eastern flanks from Spidean Coinich. The unconformity which separates these Cambrian rocks from the underlying Torridonian Sandstone can be seen from certain viewpoints. After crossing the bridge at Kylesku, stop after nearly 2 miles at a layby on the left just beyond Unapool House [NC 233316].

GEOLOGICAL VIEWPOINT: LOCH GLENCOUL

Looking east from this point provides quite the best view of a thrust fault anywhere in the British Isles. It is known as the Glencoul Thrust (MFG 147), and it affects the rocks lying underneath the Moine Thrust. This thrust is exposed on the western slopes of Beinn Aird na Loch, which is the hill with a rather broad top, lying to the east across Loch Glencoul. Its western slopes, around the promontory of Aird na Loch, consist of very rough but rather featureless ground, typical as always of the Lewisian Gneiss. These slopes lie below a very prominent line of crags which can be traced down the hillside towards the right.

These crags are formed by the Cambrian Quartzite and Pipe Rock, dipping at a low angle towards the east, and lying unconformably on top of the Lewisian Gneiss to the west. The bedding of these sedimentary rocks can easily be seen, even at a distance. These crags form an escarpment, backed by a sloping shelf that runs down to the shore of Loch Glencoul. This is a dip-slope and its eastern edge marks the outcrop of the Fucoid Beds, which are found immediately underneath the Glencoul Thrust. This plane of movement carries Lewisian Gneiss on its back, so that this formation is also exposed in typical fashion on the upper slopes of Beinn Aird na Loch to the east, overlying the rocks already described.

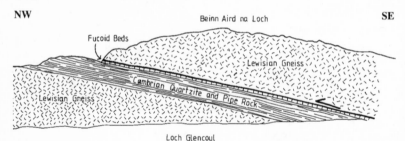

Figure 8. Vertical Cross-Section across the Glencoul Trust.

The presence of the Glencoul Thrust is clearly reflected in the topography, which shows how the Cambrian rocks are sandwiched between Lewisian Gneiss. Evidently, the Lewisian Gneiss lying on top of the Cambrian rocks has been thrust up and over these rocks from the east, probably by several miles at the very least. The Lewisian Gneiss forming the upper slopes of Beinn Aird na Loch above the Glencoul Thrust can be traced for more than 2 miles to the east towards the outcrop of the Moine Thrust at the head of Glencoul. This thrust is exposed on the Stack of Glencoul, which makes a prominent feature on the eastern skyline. It brings Moine Schists forward from the east over Cambrian Quartzite and Pipe Rock, which lies unconformably on top of the Lewisian Gneiss above the Glencoul Thrust.

GEOLOGICAL LOCALITY: LOCH GLENCOUL

A splendid exposure of the Glencoul Thrust can be visited by driving another mile south along the A894 towards Skiag Bridge. After parking just after the road crosses a bridge [NC 236303], walk east along the unconformable junction between grey Lewisian Gneiss to the north and chocolate-coloured Torridonian Sandstone to the south. After a short distance, a backward view to the west reveals the line of this unconformity as it rises to a height of 1,000 feet at the foot of Sail Garbh. Skirting round the hillside, the Cambrian Quartzite is next encountered, cutting down to the east across the Torridonian Sandstone until it comes to rest on Lewisian Gneiss at [NC 245302]. This marks the position of the double unconformity between Lewisian Gneiss, Torridonian Sandstone and Cambrian Quartzite, so characteristic of this ground. Looking back, the eye can follow the base of the Cambrian Quartzite straight to the top of Sail Garbh.

Continue down the slope around the foot of a broken escarpment, picked out by a line of small trees, at the back of Liath Bhad, where the Fucoid Beds are exposed in typical fashion. After crossing the ground to the east, scramble up a steep slope to the foot of the rocky craigs which overlook Loch Glencoul at this point [NC 259302]. These craigs are formed by Lewisian Gneiss, broken down along the line of the Glencoul Thrust into dull and very dark rocks with a platy structure. These rocks form an overhang, below which creamy-coloured dolomites of the Durness Limestone are found. There is a clean-cut line between broken-down gneiss and the underlying dolomite, which marks the plane of the Glencoul Thrust (MFG 148).

LOCH GLENCOUL TO LOCH ASSYNT

On returning to the main road, the locality described later at **Clachtoll** may now be visited by taking the B869 through Drumbeg. This scenic route joins the A894 near Unapool, less than a mile to the north. It crosses low-lying but very rough country typical of the Lewisian Gneiss along most of its course. However, south of Clashnessie Bay, it follows along the contact between these gneisses and the Torridonian Sandstone. Stop in the car-park near the beach at Clachtoll [NC 039272], just west of the road, as described below in Chapter 6.

Alternatively, the A834 can be followed south to Skiag Bridge to visit the next locality. On reaching the junction at Skiag Bridge with the A837, turn right towards Lochinver, but stop at a layby after a mile, where a straight stretch of road runs along the shore of Loch Assynt, 200 yards before some small islands with scattered pine-trees.

GEOLOGICAL LOCALITY: LOCH ASSYNT

The roadside exposures at [NC 217251] clearly show the basal beds of the Torridonian Sandstone where they rest unconformably on top of the Lewisian Gneiss. The gneiss forms the greyish and rather blocky rocks, exposed just above the road. Although rather massive in appearance, they are banded rocks, dipping moderately steeply towards the west. They are capped unconformably by dark red sandstones and shales, carrying pebbles of vein quartz, which are the well-bedded rocks exposed towards the top of the cutting. These are Torridonian in age, and the intervening surface of unconformity represents a gap in geological time of some 600 million years at the very least.

Return back along the A837 towards Skiag Bridge, crossing first from

Torridonian Sandstone on to Cambrian Quartzite at the back of a shallow valley, less than a mile from the previous locality. The Cambrian Quartzite makes a prominent escarpment, overlooking this valley, which can be traced up the slopes of Quinag towards the west. There is a marked contrast in colour between the dark Torridonian Sandstones and the light-coloured Cambrian Quartzites. After passing roadside exposures of Cambrian Quartzite, park as convenient near Skiag Bridge [NC 234244].

GEOLOGICAL EXCURSION: SKIAG BRIDGE

A short walk from Skiag Bridge, over fairly rough ground, serves to illustrate many features of the local geology. First return back along the A837 to the bend in the road at [NC 231245]. The roadcut to the north exposes typical cross-bedded Cambrian quartzites with some gritty layers. These rocks dip to the east at 10 degrees. These rocks are overlain by Pipe Rock, which is poorly exposed some 20 yards north of the road.

The view east from this point clearly shows the sedimentary sequence typical of the Cambro-Ordovician rocks. The Pipe Rock outcrops downhill as far as Skiag Bridge, where it is well-exposed in the steep face above the road. On top of this formation are the Fucoid Beds, which make the brown-weathering exposures on the hillside just above the Pipe Rock. The slight hollow running downhill towards the AA box marks the more shaly horizons that are present towards the top of the Fucoid Beds. These rocks are succeeded by the Serpulite Grit, which makes a prominent line of craigs that runs down to a promontory, jutting out into Loch Assynt opposite Ardveck Castle.

Above this escarpment there is a sharp break in slope, beyond which the Durness Limestone is exposed. Its outcrop is marked by grassy slopes, quite unlike the much wetter ground formed by the underlying rocks, covered with heather and bracken. The higher ground beyond the Durness Limestone is formed by Lewisian Gneiss, together with Cambrian Quartzite and Pipe Rock, carried forward over the underlying rocks to the west by the Glencoul Thrust. These rocks make the rocky slopes of Glas Beinn and Cnoc na Creige, lying just east of the A894 road.

Now walk uphill towards the northwest along the top of the quartzite escarpment. Crossing the stream draining from Lochan Feoir, make for a slight col at the eastern end of Lochan an Duibhe. The rocks exposed just uphill from this point [NC 223256] are pebbly Torridonian sandstones, with the bedding picked out by lines of pebbles, mostly vein quartz. The sandstone is rich in felspar, which gives the rock its pink colour. Overlying these rocks are Cambrian quartzites, exposed farther up the slope.

Walk around this slope to the north, underneath some craigs of Cambrian Quartzite, and the unconformity can be located at [NC 221258], some 150 yards north of north-east of the lochan. There, dark Torridonian sandstones are overlain by light-coloured Cambrian quartzite, dipping more steeply to the east than the underlying rocks, which are flat-lying. There is a thin pebbly grit with detrital feldspar at the base of the Cambrian sequence.

This locality provides a fine view of the double unconformity on the slopes of Beinn Garbh, beyond Loch Assynt to the south. The lower slopes of this hill are formed by rocky knolls of Lewisian Gneiss. Uphill, the slopes become steeper, and they are terraced in the manner typical of the Torridonian Sandstone. Although the Torridonian Sandstone is close to the horizontal, this formation rests unconformably on Lewisian Gneiss, burying what was once an ancient landscape of low hills and shallow valleys. The slopes of Beinn Garbh to the east are Cambrian Quartzite, dipping in the same direction at an angle close to the slope of the hillside. The quartzite cuts across the bedding of the Torridonian Sandstone towards the east, eventually coming to rest on the Lewisian Gneiss.

Clearly, the Torridonian Sandstone was tilted towards the west, prior to the deposition of the Cambrian rocks. These rocks were most likely deposited very close to the horizontal on what was a plane of marine erosion, so that they stepped across from Lewisian Gneiss in the east on to Torridonian Sandstone in the west. The Torridonian Sandstone was then tilted back towards the east, along with the overlying Cambrian rocks, so that it has now regained its original attitude, while the Cambrian rocks have come to dip towards the east.

Walking back east from this point, keep well up the slope until a path is reached north of Lochan Feoir. This leads to the wide dip-slope formed by the Cambrian Quartzite and Pipe Rock, west of Allt Sgaithaigh. Walk east over rough ground to reach this stream around [NC 230255], where it runs over a series of bedding-planes in the Pipe Rock. Excellent sections show large 'Trumpet Pipes' at this locality, showing up as large circular depressions, up to 4 inches across with a slightly raised centre, packed closely together on the bedding-surfaces. Continue east to the A894, and walk down this road to Skiag Bridge.

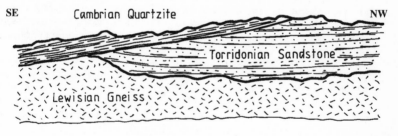

Figure 9. The Double Unconformity south of Loch Assynt.

The road follows the contact between the Pipe Rock and the Fucoid Beds south towards Skiag Bridge. After passing exposures of the Fucoid Beds, the Pipe Rock can be examined in the steep rock-face, close to the road junction. It consists of purplish and white quartzites, cross-bedded in places and cut by a multitude of vertical pipes, up to 18 inches in length and filled with much whiter quartzite. These pipes are the traces of worm-burrows, filled with fine sand. They give a typically pitted appearance to the bedding-surfaces, which often allows the Pipe Rock to be recognised in the field.

The contact between the Pipe Rock and the overlying Fucoid Beds can be found by walking 200 yards east along the A837 towards Inchnadamph. Unlike the light-coloured quartzites forming the uppermost beds of Pipe Rock, the Fucoid Beds are brown-weathering sandy dolomites, with rather irregular bedding and much shaly material. The lowermost beds of this formation are seen at the roadside, while there are also good exposures along the shore of Loch Assynt, opposite the AA box.

Careful examination of the bedding-planes exposed below the road may reveal worm-like traces, up to half an inch across, which are the flattened remains of organic burrows, filled with sand. It was these markings that the early geologists thought were made by seaweed, so accounting for the name given to these beds. The uppermost horizons of the Fucoid Beds are mostly shales, which outcrop along the hollow which runs uphill from the AA box, just to the west of the crag formed by the Serpulite Grit.

The Serpulite Grit can next be examined along the roadside to the east of the AA box, where it forms a gritty quartzite, often somewhat calcareous to judge by its weathering. It resembles the underlying Pipe Rock in that worm burrows are present as vertical pipes, filled with sand. The fossils that gave this horizon its name are now known as Salterella. They are preserved abundantly on some bedding-planes, forming conical, worm-like impressions, up to a quarter inch in length. They are best seen where the Serpulite Grit reaches Loch Assynt in a low promontory, across the bay from Ardvreck Castle.

Return to the road where the lowermost beds of the Durness Limestone are seen to rest abruptly on top of the Serpulite Grit, some 50 yards to the east of the AA box. These dark limestones share the same easterly dip as the underlying rocks. However, walking along the road to the east reveals that a sudden change in dip at the back of the small bay, north of Ardvreck Castle. This marks the trace of the lowermost thrust to affect the rocks of the Assynt region, well in advance of the Moine Thrust, which lies several miles farther east at this point.

SKIAG BRIDGE TO KNOCKAN CLIFF

The route south from Skiag Bridge along the A837 runs past Inchnadamph along the valley of the River Loanan. It follows the outcrop of the lowermost thrusts to affect the Moine Thrust Belt to the east. The Sole Thrust itself occurs within the Fucoid Beds along most of its course, although it occasionally rises into the Durness Limestone, close to its base. The Basal Quartzite and Pipe Rock are therefore not affected by the thrusting. They are exposed west of the road in very long dip-slopes, descending east from the summits of Beinn Garbh, Canisp, and Cul Mor.

The Cambrian Quartzite and Pipe Rock are highly impermeable rocks, giving rise to very marshy ground covered in peat and heather, while the Durness Limestone to the east often weathers into a good soil, covered with grass. Just before Inchnadamph, the A837 passes a memorial to Peach and Horne, who mapped this ground, and indeed much of the Northwest Highlands, as officers of the Geological Survey of Scotland during the final decades of the 19th century. It lies west of the road at the top of a small hill.

A short walk up the valley of the *River Traligill* from Inchnadamph provides an excellent view of a thrust-plane, where it is exposed in the river-bed at

[NC 267212], southeast of Glenbain Cottage. Upstream, the river runs below ground for a quarter of a mile, and the best exposure of the thrust-plane is found where it reappears at the surface. Such underground drainage is typical of rivers in limestone terrain, where the solid rock can be dissolved away by slightly acid waters.

Just south of Inchnadamph, the road runs for well over a mile below the escarpment of *Stronrubie*, formed by the Durness Limestone, which displays imbricate structure typical of rocks affected by thrust-faulting. Beyond this point, however, there is little of geological interest until the hills of Cnoc an Leathaid come into view, west of Loch Awe.

The Sole Thrust advances to the west at this point, carrying forward Lewisian Gneiss which, with its cover of Cambrian Quartzite and Pipe Rock, makes the summits of these two hills. East of the road, the upper slopes of Beinn nan Cnaimhseag and Beinn an Fhuarain expose Torridonian Sandstone, thrust westwards over the underlying Durness Limestone on the Ben More Thrust.

Turn right at Ledmore along the A835 towards Elphin, where the Durness Limestone is thrust over the Fucoid Beds and Serpulite Grit. The limestone is well-exposed in the cliffs overlooking the village. After crossing the Durness Limestone, the road passes over Serpulite Grit and the Fucoid Beds as it ascends the hill to the south. Just beyond the cairn marking the boundary of the Inverpolly Nature Reserve, turn left into the car-park at the foot of Knockan Cliff [NC 188091].

GEOLOGICAL LOCALITY: KNOCKAN CLIFF

Knockan Cliff is a National Nature Reserve, so declared for its geological importance by Scottish Natural Heritage. It provides quite the best exposure of the Moine Thrust anywhere in the Northwest Highlands. The car-park looks out over a dip-slope of Cambrian Quartzite, descending towards the road from the slopes of Cul Mor to the northwest. Beyond lies a terraced landscape of Torridonian Sandstone, building the nearby mountains of Cul Mor and Cul Beag, while Stac Pollaidh is seen on the far skyline, beyond Gleann Laoigh.

Walk up the path from the car-park, passing the Pipe Rock in the quarry to the left, and then crossing the Fucoid Beds and the Serpulite Grit as the path climbs up the hillside. Just below the dark crags at the top of this slope, creamy-white dolomites are exposed, marking the outcrop of the Durness Limestone. These dolomites are separated from the rocks below by the Sole Thrust, which is not exposed. There is then an abrupt contact between these dolomites and much darker rocks which make the cliffs at the top of the slope. These dark rocks are the Moine Schists, and their abrupt contact with the underlying rocks is the line of the Moine Thrust (MFG 149).

This thrust can be traced for some 200 yards along the face of the cliff, everywhere forming a sharp break between light-coloured dolomites and much darker schists. In some places, the weathering of the dolomite has produced an overhang along this contact, where it can be seen that the Moine Thrust is inclined towards the east at quite a low angle, 10 degrees at the very most. The rocks lying just above the Moine Thrust are not typical of the Moine Schists as a whole. The vast movements on this thrust-plane have reduced these schists to what are mylonites all along its course, breaking down and drawing out the original constituents in the rock.

On reaching the top of the slope north of Knockan Cliff, walk northeast for nearly a mile over Durness Limestone to an excellent viewpoint above Elphin. The Torridonian mountains of Suilven and Canisp are conspicuous to the north, capped by Cambrian Quartzite on Canisp. Quartzite and Pipe Rock then form a long dip-slope, descending towards the east. It ends in the outcrop of the Sole Thrust, following the Fucoid Beds or the Durness Limestone, across the lower ground towards Inchnadamph, as already described. The thrust mass of Lewisian Gneiss, capped by Cambrian rocks, forms the prominent hills of Cnoc an Leathaid, just north of Cam Loch. Farther east, Torridonian Sandstone forms an isolated mass on Beinn an Fhuarain, thrust over the underlying Durness Limestone. Cnoc na Sroine above Loch Borolan is syenite, forming an igneous intrusion into the surrounding rocks.

The higher and more distant hills on the skyline to the northeast, including Sgonnan Mhor and Ben More Assynt, are mostly Lewisian Gneiss, thrust together with Cambrian Quartzite and Pipe Rock over the underlying rocks to the west. The Moine Thrust does not appear at all in this view, since it makes a wide detour away from its normal course, northeast of Knockan Cliff, eventually passing to the east of Ben More Assynt. Only north of the Stack of Glencoul does it regain its original course. Return to the car-park from the view-point, by the same route, avoiding the steep slopes above the road.

KNOCKAN CLIFF TO ULLAPOOL

Continue south along the A835 towards Ullapool, following the outcrop of the Moine Thrust for most of the way, just east of the road. The many cuttings along this road expose flaggy mylonites at first, before the road junction to Achiltibuie is reached. This unclassified road may be followed to reached the locality of *Enerd Bay*, which is described in the next chapter. South of this junction, there are roadside exposures along the A837 of Cambrian Quartzite and Pipe Rock, while Ben More Coigach makes a prominent mountain of Torridonian Sandstone to the west.

After crossing Strath Kanaird, the road runs along a narrow valley, eroded along the line of a fault. It separates Cambrian Quartzite to the northwest from Torridonian Sandstone to the southeast. The quartzites in particular have been shattered by the faulting. The road then crosses Torridonian Sandstone, exposed in typical fashion, before it enters Ullapool. This village is built upon a wide terrace of superficial deposits, extending out into Loch Broom.

CHAPTER 7

Clachtoll and Enard Bay

CLACHTOLL MAY BE REACHED most easily by taking the A837 road from Skiag Bridge towards Lochinver. On reaching the outskirts of this village, turn right along the B869 towards Clachtoll, and stop in the car-park, just west of the road at [NC 039272]. Alternatively, the B869 may be followed through Drumbeg, after leaving the A894 just south of Unapool.

GEOLOGICAL LOCALITY: CLACHTOLL

The unconformity between the Torridonian sandstones of the older Stoer Group and the underlying Lewisian Gneiss can first be examined in the crags just east of the road. Cross the road from the car-park, and walk up a track past a small quarry. This exposes red Torridonian mudstones with well-marked jointing, which pass downwards into well-bedded grits and breccias to the east. After passing a cottage, climb up a broad gully to the top of a small hill overlooking the car-park.

Its top is formed by a glacial pavement, exposing coarse breccias with many fragments of acid gneiss, together with scattered blocks of a dark ultra-basic rock. Behind this hill is a ruined croft [NC 043272], and the unconformity can be located just a few yards to its west, where coarse breccias rest directly on top of the Lewisian Gneiss. The unconformity can then be followed south towards the road, although it is nowhere exposed, separating breccias to the west from gneisses to the east.

Cross the road, and walk south across the fields, keeping well to the left of the jagged rocks guarding the entrance to Clachtoll Bay. A faulted contact between Torridonian Sandstone and the underlying gneiss is hidden by sand at a small inlet on the south coast at [NC 039268]. The Lewisian Gneiss forming the rock-face immediately east of this inlet is cut by a series of Neptunian dykes (MFG 104), filled with chocolate-coloured mudstone.

These fissures opened up in Torridonian times, allowing sediment to be washed in from the surface. Angular fragments of the gneiss, broken off the walls, lie surrounded by mudstone in some places, and locally the rock passes into a breccia. Beyond these exposures to the southeast, there is a tongue of Torridonian breccia, just above high-water mark, plastered against a cliff of Lewisian Gneiss.

Returning to the sandy inlet, red mudstones are exposed to the west beyond the Lewisian Gneiss. Occasional beds of gritty sandstone are present in these mudstones, and the lowermost sandstone is cut by two sets of thrust-faults, dipping at low angles towards one another. The mudstones are cut by a complex network of very closely-spaced joints, which obscures the bedding in these rocks. Walking back around the coast towards Clachtoll, the unconformity can be located again just before the path descends to the beach at [NC 040270].

The Torridonian rocks of the headland are cut off by a fault, which brings up Lewisian Gneiss so that it is exposed along the coast just southwest of the beach. The banding of this gneiss dips steeply towards the southwest. Mudstones can be seen resting against this gneiss just to its northeast, forming a near-vertical unconformity, along which a thin breccia is found. This contact follows the banding in the gneiss, and evidently formed a low cliff, against which the Torridonian rocks were deposited.

Walk northwest around the Bay of Clachtoll to the west coast just south of Sgeir na Traghad at [NC 036274]. The mudstones exposed around the Bay of Clachtoll give way to cross-bedded sandstones in this direction, forming slightly higher ground to the west. These sandstones are extremely well-exposed along the west coast, where they occur as thick beds separated from one another by shales and mudstones. Dipping out to sea at 30 degrees, there is one bedding-plane, exposed over a very wide area. It is crossed by the trace of a normal fault, which forms an irregular step in the bedding (MFG 124).

CLACHTOLL TO ENARD BAY

Return to the car park, and drive southeast towards Lochinver along the B869 across a rocky landscape, again typical of the Lewisian Gneiss. The Ordnance Survey Map (Sheet 15) shows the road running along a narrow steep-sided valley, which can be traced towards the southeast from Maiden Loch, past Rhicarn and Brackloch, and across the River Inver. It marks the trace of an ultrabasic dyke, which is particularly susceptible to weathering and erosion. There are many such features cutting across the Lewisian Gneiss in this direction, imparting a pronounced grain to the country.

After driving up the steep hill near Polla, stop at the summit [NC 075257], where there is a fine view of Suilven, Canisp and Cul Mor, rising abruptly as isolated relics of Torridonian Sandstone above a low-lying but very rough platform of Lewisian Gneiss. The view is even better from the slightly higher ground just north of the road.

On reaching the A837, turn right and drive through Lochinver. Take the unclassified road south past the school towards Inverkirkaig and beyond. The twists and turns of this single-track road are typical of the country formed by the Lewisian Gneiss, which may be examined almost anywhere along its course.

After crossing the River Polly, there is an abrupt change in topography as the road climbs over the Aird of Coigach. The extremely rough country of rocky knolls, lochans and peaty hollows, characteristic of the Lewisian Gneiss, gives way to rolling moorlands covered with peat, formed by the Torridonian Sandstone, mantled by boulder clay on the lower ground.

There is a fine view from the summit of the Aird of Coigach [NC 074115], dominated by the sharp peak of Stac Pollaidh just to the east. All the mountains are Torridonian Sandstone, ranging from Ben More Coigach to the southeast, Cul Mor to the east, and Suilven and Canisp to the northeast. On reaching Loch Bad a'Ghail, turn right towards Achiltibuie, and park at a bend in the road at [NC 025127]. As already mentioned, it is possible to reach this locality after visiting Knockan Crag.

GEOLOGICAL LOCALITY: ENARD BAY

Walk north over easy ground towards the coast east of Achnahaird Bay. A rocky knoll, scoured smooth by the ice, exposes Lewisian Gneiss surrounded by Torridonian Sandstone at [NC 023135]. Farther on, coastal exposures around [NC 022143] consist of red Torridonian sandstones with cross-bedding in large units, up to 12 feet thick. The internal bedding within these cross-bedded units is very well-exposed as curving surfaces, dipping more steeply than the bedding itself. The currents depositing these sandstones evidently flowed from the northeast. Farther north, beyond a small bay, which marks the trace of a fault, climb over the shoulder of Cnoc Mor an Rubha Bhig to reach the north coast at a narrow inlet [NC 027143]. This inlet lies below a steep slope of Torridonian Sandstone belonging to the Stoer Group, down which a sheep track can be followed to the shore.

These sandstones are exposed as a slight overhang along the western wall of the inlet. Dipping moderately towards the west, they rest with an erosive contact on top of red mudstones at the back of the inlet. These mudstones are very much jointed and fractured. They become gritty as they are traced to the east, where they pass over a distance of a few yards into a gneiss breccia.

Walk to the seaward end of a slight promontory which lies just east of this inlet, where these breccias are best exposed, facing out to sea. They are banked against a hill of Lewisian Gneiss which forms the promontory itself. The contact is close to the vertical along its western side. More breccias are found around the eastern side of this promontory, passing into gritty mudstones, banked against the Lewisian Gneiss. All these exposures lie to the west of an almost enclosed bay, which can be identified by the ruins of a bothy at [NC 028146].

Walk around the shores of this bay past the ruined bothy to the next headland, passing exposures of Torridonian conglomerate. The seaward end of this headland is formed by another hill of Lewisian Gneiss, mantled by coarse breccias carrying angular fragments of the underlying gneiss. These breccias are well-exposed along the western side of the headland. They can be traced eastwards across a narrow neck of land where the sea has almost cut off this headland, about 100 yards north of the bothy. There, the gneiss breccia has a matrix of fine-grained limestone, which weathers away to form a fretted surface, typical of calcareous rocks.

The breccia then passes up into a thin horizon of finely-laminated limestone at its very top, carrying scattered blocks of gneiss. This limestone underlies a few feet of red mudstones, forming an easily-eroded horizon penetrated to some extent by the sea. All these rocks form a mound, over which Torridonian conglomerate appears draped to the south. This conglomerate is quite unlike the underlying breccias with fragments of Lewisian Gneiss because it carries large boulders of sandstone, thought to be derived from the sandstone outcrop just to the west.

This sandstone conglomerate is exposed along the coast towards the east, while its base can be traced inland to where it reaches the coast again, just west of the next bay to the east [NC 030147]. There, the conglomerate is underlain by red mudstones, which carry thin beds of gritty limestone. The contact has been undercut by the sea. The mudstones pass down into drab red sandstones, which form slabby exposures on the foreshore.

At a slight headland, 100 yards to the east of this contact, careful search will reveal accretionary lapilli as pea-sized nodules, exposed on the bedding. These are thought to be fossil rain-drops, or possibly hail-stones, formed by the accretion of volcanic dust in the rain-clouds that accompany volcanic eruptions. Walk south to the road from this point, over ground which provides excellent views of the Torridonian mountains towards the east beyond the shores of Enard Bay.

CHAPTER 8

Ullapool to Kyle of Lochalsh

THE GEOLOGY DOES NOT CHANGE greatly south of Ullapool, as all the features recognised farther north can be traced from Assynt into Wester Ross. The Moine Thrust with its belt of structural complication is still present, while the Moine Schists are exposed widely in the ground lying farther to the east. Northwest of the Moine Thrust lie the Lewisian Gneiss, Torridonian Sandstone and Cambro-Ordovician rocks of the Caledonian foreland, where the Caledonian earth-movements affecting the Moine Schists had little or no effect.

Again, the foundations of the geology determine the nature of the landscape. The Lewisian Gneiss is far more restricted in its outcrop, so that the extremely rough terrain typical of these rocks in the Northwest Highlands is absent farther south, except locally around Gairloch and Torridon. The Torridonian Sandstone outcrops much more widely, exposed in a magnificent series of true mountains with great corries and rocky ridges, like An Teallach, Beinn Eighe, Liathach, and the Applecross hills, quite unlike the Torridonian hills farther north. Equally, the Moine Thrust does not make such a prominent feature in the landscape, except locally, as it does not carry forward the great masses of Lewisian Gneiss, which make up mountains like Ben More Assynt along its course farther to the north.

ULLAPOOL TO DUNDONELL

Starting at Ullapool, drive east along the A835 towards the head of Loch Broom. The road passes over Torridonian Sandstone on to Cambrian Quartzite and Pipe Rock, after which the Fucoid Beds are encountered, weathering rusty-brown in typical fashion. The road then crosses the Sole Thrust, exposed just east of a bridge, 1½ miles from Ullapool, where Torridonian Sandstone and Cambrian Quartzite are thrust over Durness Limestone.

Just after the road returns to the coast beyond *Corrie Point*, mylonites (MFG 131) are exposed in a series of roadcuts. These are very black and extremely fine-grained rocks, in which folding can be seen where it affects very thin but continuous bands of lighter-coloured rock. These intensely deformed rocks, lacking any trace of original character, occur just above the Moine Thrust at this locality.

The trace of this thrust can be seen across Loch Broom, where it follows the break between the craggy ground underlain by Torridonian Sandstone (and Lewisian Gneiss) to the west, and the much smoother terrain to the east. Continuing towards the east, the Moine Schists lying above this thrust are clearly exposed on the hillsides above Loch Broom, mostly dipping eastwards at a gentle angle. These rocks are also seen in a series of roadside exposures after passing the head of Loch Broom.

Stop in the car-park for *Corrieshalloch Gorge,* a mile beyond Braemore. It is a very deep but narrow gorge, eroded by glacial melt-waters as the ice retreated at the end of the Great Ice Age. The view from the suspension bridge over the Falls of Measach shows flat-lying Moine Schists, cut by very prominent joints in a northwesterly direction. These joints evidently acted as planes of weakness, along which the gorge itself was eroded. It is nearly a mile long, and 200 feet deep. The vertical walls follow the joint-planes in the rock.

Return to the car-park, continue east along the A835 for half a mile, and then turn right at the road junction along the A832 towards Dundonell. The route first passes over Moine Schists for several miles. Crossing the high ground towards Dundonell, there are excellent views northwest towards An Teallach. Descending towards the north, the road enters the deep valley of the Dundonell River, overlooked on its east by steep crags of Moine Schists.

The road crosses the Moine Thrust at the mouth of this valley, where it opens out into Strath Beag. The dip-slope facing the road at this point is Cambrian Quartzite, making a very pronounced feature running across the southeastern slopes of An Teallach. Cambrian Quartzite also occurs as outliers on the An Teallach ridge and its eastern spurs, capping the tops of Sail Liath, Glas Mheall Liath and Glas Mheall Mor. The quartzite rests unconformably on Torridonian Sandstone, making up the other summits of An Teallach, and magnificently exposed in the precipices of its eastern corries.

Stop at a layby [NH 108868], half a mile to the north of Dundonell House. The view east shows a cross-section through the Moine Thrust. Creag Chorcurach is Moine Schist, resting on top of the thrust, which lies hidden under the screes lower down the slope. The rocks below the thrust are Cambrian Quartzite, forming the low crags farther to the west. Beyond, Torridonian Sandstone makes the lower ground before rising towards the summit of Beinn nam Ban.

DUNDONELL TO LOCH MAREE

The road continues northwest from Dundonell along the shores of Little Loch Broom, passing through a landscape typical of the Torridonian Sandstone until it reaches Mungasdale on the eastern shore of Gruinard Bay. There is then an abrupt change in topography as the unconformity is crossed from the Torridonian Sandstone on to the underlying Lewisian Gneiss.

The gneiss is exposed in a great many rocky knolls and crags, separated from one another by steep-sided valleys. Typical exposures are seen west of the Little Gruinard River where the road climbs a steep hill to the viewpoint at [NH 940901]. The Lewisian Gneiss at this point forms an intimate mixture of acid and basic gneiss, where veins of light-coloured granite have invaded the much darker basic gneisses in a very irregular manner.

Beyond this point, the road crosses back on to the Torridonian Sandstone, which makes the much low ground mantled with boulder clay around the shores of Loch Ewe. Lewisian Gneiss is encountered once more just beyond Poolewe, where it is brought to the surface by a fault that runs northwest from Loch Maree along the course of the River Ewe. On crossing this river at Poolewe, there is a wide and very flat terrace of fluvio-glacial deposits around Pool Crofts.

This terrace is backed by high ground where the Lewisian Gneiss is exposed in an escarpment which follows the line of the Loch Maree Fault.

Just over a mile south of Poolewe, there is an excellent view from the road towards Loch Maree, looking southeast from the viewpoint above Tollie Farm. The road then continues across Lewisian Gneiss towards Gairloch, where the Torridonian Sandstone reaches the coast. The village is built on another terrace at a height of some 50 feet above sea-level.

Continue south from Gairloch along the A832, crossing over Lewisian Gneiss until the shores of Loch Maree are reached around Talladale. The Torridonian Sandstone is encountered once more beyond this point, as shown by the terraced nature of the hillsides. Good exposures are seen along the road. Looking across Loch Maree, it can be seen that Lewisian Gneiss outcrops along its northeastern shore, which follows the straight northwesterly trend of the Loch Maree Fault. Park at the layby [NG 982670], just over a mile past the Bridge of Grudie.

GEOLOGICAL PANORAMA: LOCH MAREE

The well-known view across Loch Maree from this point shows the shapely peak of Slioch, rising steeply to the northeast. Its foundations are Lewisian Gneiss, which can be identified even at a distance by the silvery-grey outcrops on the lower slopes of this mountain. The gneiss is covered unconformably by well-bedded Torridonian Sandstone, typically dark-purple in colour, making the upper ramparts of this mountain. This unconformity represents a very ancient land-surface of hills and valleys in the Lewisian Gneiss, which were subsequently buried underneath the Torridonian Sandstone. This is well seen around Meall Riabhach, where the Lewisian Gneiss clearly makes quite a large hill of pale-coloured rock, surrounded by Torridonian Sandstone, against which the sandstone has been deposited as scree deposits.

LOCH MAREE TO LOCH TORRIDON

Continue along the A832 towards Kinlochewe. After a mile, Torridonian Sandstone gives way to Cambrian Quartzite and Pipe Rock, overlain in some places by Fucoid Beds and Serpulite Grit, all dipping towards the east at a low angle. They form the lower slopes of Meall a'Ghiubbhais around Coille na Glasleitre. However, the summit of this hill is Torridonian Sandstone, which occurs as an outlier, thrust into this position over the underlying Cambrian rocks.

Viewed from a distance, the Torridonian Sandston forming the summit of Meall a'Ghiubbhais appears as a dark mass, fringed by light-coloured rocks. Farther along the road towards Kinlochewe, the Cambrian rocks come to dip much more steeply towards the southeast, as they are increasingly affected by folding and thrusting in this direction. This is again a consequence of the Caledonian earth-movements associated with the Moine Thrust, which passes just to the southeast of Kinlochewe.

There is an excellent view across Loch Maree from the car-park at the foot of Meall a'Ghiubhais. The deep valley of Gleann Bianasdail makes a prominent feature to the northeast, between Slioch and Beinn a'Mhuinidh, eroded along a fault. This throws Torridonian Sandstone down to the southeast against Lewisian Gneiss to the northwest. The prominent escarpment southeast of this valley is Cambrian Quartzite and Pipe Rock, lying unconformably on top of Torridonian Sandstone which forms the lower slopes of Beinn a'Mhuinidh.

Overlying the Cambrian rocks of this escarpment comes Lewisian Gneiss, exposed widely farther north around the summit of Beinn a'Mhuinidh. This has been thrust over the underlying Cambrian rocks, along with Torridonian Sandstone. The effects of these earth-movements can be seen just to the southeast on the slopes above Kinlochewe, where it is clear that the rocks are much disturbed.

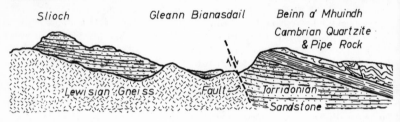

Figure 10. Vertical Cross-Section to the North of Loch Maree.

Turn right at Kinlochewe along the A896 towards Loch Torridon. The road follows a wide outcrop of Torridonian Sandstone, sandwiched between Cambrian Pipe Rock to the northwest and the Moine Thrust to the southeast. Near Loch Clair, the road crosses these Cambrian rocks to reach another outcrop of Torridonian Sandstone, lying well beyond the Moine Thrust. The Cambrian Quartzite and Pipe Rock can be traced from Loch Clair up the eastern slopes of Beinn Eighe, where they are much disturbed by folding and thrusting, along with the Torridonian Sandstone. The greater part of the ridge forming Beinn Eighe is Cambrian Quartzite and Pipe Rock, folded and thrust in a complex fashion. Only at its western end, around Sail Mhor, do these rocks revert to a horizontal attitude, unaffected by the Caledonian earth-movements.

South of the road beyond Loch Clair lies Sgurr Dubh, which consists of steeply-dipping Torridonian Sandstone and Cambrian Quartzite. Slightly farther west, Cambrian Quartzite forms a prominent line of light-coloured crags on the northwestern slopes of this mountain, surrounded by much darker Torridonian Sandstone. The lower ground below these crags is covered by a wide expanse of hummocky moraine, particularly to the southwest of Lochan an Iasgair.

Continuing west into Glen Torridon, the road runs along the foot of Liathach, another mountain of Torridonian Sandstone, capped by Cambrian Quartzite on several of its peaks. The Torridonian Sandstone shows easterly dips below Stuc a'Choire Dhuibh Bhig, the most easterly of its peaks, but these rocks become flat-lying farther west, after a fault is crossed. The southern slopes of Liathach are virtually all rock beyond this point, exposing the Torridonian Sandstone splendidly in steep tiers of rocky terraces above the road.

Turn right at the foot of Glen Torridon along the minor road towards Diabaig along the northern shores of Loch Torridon. After passing Inveralligin, the road climbs over Bealach na Gaoithe, where a stop can be made at a layby [NG 827861], 200 yards uphill from the hairpin bend.

GEOLOGICAL PANORAMA: LOCH TORRIDON

The rocks in the immediate vicinity of this viewpoint are Lewisian Gneiss, which has been faulted against the Torridonian Sandstone, exposed to the east beyond the valley of Abhainn Alligin. This steep-sided valley is eroded along the line of a fault, and there is an abrupt change in the landscape to the west, where the Lewisian Gneiss is typically exposed in a series of low rocky knolls. Looking south across Loch Torridon, the unconformity between the Torridonian Sandstone and the Lewisian Gneiss can clearly be seen on a good day. The gneiss can be recognised even at this distance, where it forms the silvery-grey outcrops along the coast. It is overlain by much darker Torridonian Sandstone, clearly displaying its well-bedded character.

Although flat-lying, the Torridonian Sandstone rests unconformably on top of an undulating surface of Lewisian Gneiss, burying this ancient landscape. The Lewisian Gneiss now forms a ridge which ends in the promontory north of Shieldaig. Torridonian Sandstones lies just to the east, where it is exposed on the slopes just behind the almost enclosed bay of Ob Mheallaidh. Beyond this point, the Lewisian Gneiss rises to a height of 800 feet on the eastern slopes of Beinn Sheildaig, only to descend nearly to sea-level once more around Balgy.

LOCH TORRIDON TO KYLE OF LOCHALSH

Return to the main road at the head of Loch Torridon, and turn right along the A896 towards Shieldaig. The pre-Torridonian hill of Lewisian Gneiss can be seen at closer quarters, just beyond Balgy. From Shieldaig, the road runs inland along Glen Shieldaig towards the Moine Thrust, crossing over Torridonian Sandstone. Stop just beyond the head of this valley, where there is a fine view over the open country to the southeast.

The lowermost slopes beyond Strath a'Bhathaich expose Cambrian Quartzite and Pipe Rock, forming a slight escarpment that runs southwest towards Loch an Loin. The flat-lying ground below this escarpment is Torridonian Sandstone. The Durness Limestone outcrops over a wide area above this escarpment, forming the pale-coloured exposures on the lower slopes of Sgurr a'Gharaidh. It has been thrust over the underlying rocks from the southeast.

Another thrust then brings forward Lewisian Gneiss on top of the Durness Limestone, resulting in a very sharp contact towards the top of Sgurr a'Gharaidh between pale limestone and the very dark gneiss on top. This marks the line of the Kishorn Thrust, which makes an important feature of the local geology as far southwest as Skye. Looking northeast, the hills to the northwest of Strath a'Bhathaich are Torridonian Sandstone and Cambrian Quartzite, folded and thrust together by the Caledonian earth-movements.

Continue south along the road towards Kishorn. West of this road, Torridonian Sandstone is magnificently exposed in the precipices of Beinn Bhan and Sgurr a'Ghaorachain. Exposures of Durness Limestone are seen at the roadside, south of Couldoran. These outcrops continue as far south as Sanachan, where the road turns inland.

The rocks beyond this point are Torridonian Sandstone, lying above the Kishorn Thrust. They have been turned upside down by the Caledonian earth-movements. They now dip moderately towards the southeast as the road ascends the valley of Abhainn Gumhang a'Ghlinne towards the crags at its head. They consist of Lewisian Gneiss, overlying the Torridonian Sandstone farther west. The Moine Thrust is crossed on the descent towards Lochcarron, while roadcuts expose flaggy mylonites just before this village is reached.

Turn left at Lochcarron towards Achnasheen, following the A896 to the head of Loch Carron. Turn right at the road junction with the A860, and continue along the A890 through Strathcarron towards Kyle of Lochalsh. The steep hillsides southeast of Loch Carron are Lewisian Gneiss, thrust over a narrow strip of Moine Schist along the coast. Beyond Stromeferry, a detour can be made through the picturesque village of Plockton by turning right at Achmore, and following the signposts. Drive through Plockton to the viewpoint at Rubha Mor [NG 807342].

GEOLOGICAL PANORAMA: RUBHA MOR

The view to the northeast displays a 'remarkable piece of topography, without parallel in the Northwest Highlands', where the Lewisian Gneiss makes a series of bold and very conspicuous escarpments on both sides of Loch Carron. These can be traced from An Sgurr and Bad a'Chreamha in the north, through Creag Mhaol and Creag an Duilisg just across Loch Carron, to Carn a'Bhealaich Mhoir and Carn an Reidh bhric in the south. Throughout this area, the Lewisian Gneiss has been turned upside down through nearly 180 degrees so that it now rests on top of the Torridonian Sandstone, itself inverted by the very same movements.

The Torridonian Sandstone forms the lower ground to the northwest, overlooked by the Lewisian escarpment, while the contact between gneiss and sandstone dips underneath this escarpment at a shallow angle towards the southeast. The escarpment continues farther south, broken by faulting, but the Lewisian Gneiss is now thrust over the Torridonian Sandstone, itself still inverted throughout this area. This high ground runs from Letter Hill and Carn Greannach in the north, towards Auchtertyre Hill and Sgurr Mor, overlooking Loch Alsh to the south.

To reach Kyle of Lochalsh from Plockton, take the back road that runs along the coast through Duirinish, crossing the outcrop of the Torridonian Sandstone. Fine views are seen towards Applecross and the Isles of Skye.

CHAPTER 9

Isle of Skye

THE TRAIL NOW LEAVES THE Moine Thrust and its ancient foreland for the Tertiary volcanic rocks on the Isle of Skye. However, the Caledonian earth-movements still affect the rocks around Sleat in the southeast of the island. The Moine Thrust crosses the Sound of Sleat, giving rise to a narrow outcrop of Lewisian Gneiss along the southeastern coast of this peninsula. Beyond, there is a wide outcrop of Torridonian Sandstone to the northwest, lying on top of the Kishorn Thrust. These rocks have been carried over the Cambro-Ordovician rocks, exposed around Ord and farther north, between Broadford and Torrin.

Elsewhere on Skye, igneous rocks of Tertiary age are mostly exposed. A thick sequence of basalt lavas is found over a wide area in the north of the island, forming the peninsulas of Trotternish, Waternish and Duirinish. This ground forms a lava-plateau, rising to more than 2,000 feet in places, and falling away steeply towards the sea in great cliffs. The step-like breaks in this landscape, known as trap features, are formed by the differential erosion of the separate lava-flows which make up this sequence of volcanic rocks. Triassic and Jurassic rocks are preserved below the Tertiary lavas, in the south around Broadford and Elgol, and farther north beyond Portree.

The scenery of Skye is dominated by the igneous rocks of the Red Hills and the Black Cuillins, standing high above the basalt country to the north and the pre-Tertiary rocks to the south. These rocks are the intrusive roots of Tertiary volcanoes, now greatly reduced by erosion. The Red Hills are mostly granite, which outcrops as several distinct intrusions between Broadford and Sligachan. The Black Cuillins are mostly gabbro, quite unlike the Red Hills to their east. Together with Blaven (Bla Bheinn on the Ordnance Survey Map), these gabbroic mountains are among the most rugged in the British Isles, best explored only by experienced climbers and hill-walkers.

KYLEAKIN TO BROADFORD

After crossing the toll-bridge from Kyle of Lochalsh, follow the A850 towards Broadford, crossing at first over Torridonian Sandstone for several miles. A raised beach runs along the coast at a height of 100 feet, and its deposits of sand and gravel are quarried. Jurassic rocks are then encountered after the road junction to Kylerhea, and these sedimentary rocks make the low ground around the straggling village of Broadford itself. They consist of well-bedded limestones, sandstones and shales, rich in fossils. The lowermost beds can be studied along the coast just west of *Ob Lusa* [NG 700248], while the higher beds are reasonably well-exposed farther to the west along the northwestern shore of *Ardnish*. Low tide is needed for all these exposures.

SOUTH OF BROADFORD

The Sleat peninsula south of Broadford lies within the Moine Thrust Belt, where it continues towards the southwest from the mainland. The Moine Thrust is exposed along the southeast coast of Sleat, bringing Lewisian Gneiss forward from the southeast to rest on top of Torridonian Sandstone. Apart from very complex geology around the Point of Sleat, other features of geological interest are seen around Ord, where Cambrian Quartzite and Durness Limestone come up from depth to outcrop at the surface, surrounded by Torridonian Sandstone. As this is the exact opposite of what would normally be expected, most likely the Torridonian Sandstone had been already thrust over the Cambro-Ordovician rocks by the Kishorn Thrust, before all these rocks were folded.

Take the A851 from Skulamus near Broadford, driving south towards Armadale across Torridonian Sandstone. Jurassic rocks outcrop west of the road, where they dip at a shallow angle towards the north, as shown by a series of dip-and-scarp features in the landscape. After reaching the shore of Loch na Dal, the road crosses the Moine Thrust at Camas nam Muilt (see OS Sheet 33), bringing Lewisian Gneiss forward on top of Torridonian Sandstone to the northwest. The gneiss makes rather bolder crags than the sandstones, even although it gives rise to more fertile ground, covered in grass rather than heather. It makes a narrow strip of land, up to 2 miles wide, overlooking the Sound of Sleat to the southeast.

Continue along the A851 towards Armadale, and turn right along the road towards Tarskavaig, south of Kilmore. This road crosses back on to Torridonian Sandstone after a short distance. Turn north on reaching the west coast at *Achnaclioch*, where a Tertiary dyke is well-exposed. Exposures of red sandstones and grey shales are seen along the road, just beyond Tokavaig, representing the Torridonian Sandstone.

North of this point, the road is overlooked by outcrops of Cambrian Quartzite, forming the bare slopes and rather rounded summits of Sgiath-bheinn Tokavaig, Sgiath-bheinn Crossavaig, Meall Da-bheinn, and Sgiath-bheinn an Uird. The whiteness of the quartzite makes a striking contrast with the drab nature of the Torridonian Sandstone. This quartzite ridge, rarely more than 900 feet high, is flanked to the west by Durness Limestone, which outcrops on the lower slopes just above the road. There is a fault between this limestone and the Torridonian Sandstone farther to the west, which the road follows as it approaches Ord.

The road turns inland at Ord, first crossing over outcrops of Durness Limestone for a few hundred yards before it reaches the Cambrian Quartzite. The junction between limestone and quartzite follows a slight valley, along which Fucoid Beds and Serpulite Grit are exposed in a few places, together with remnants of Pipe Rock. Where exposed south of Cnoc na Fuarachad, the quartzite lying to the east of this valley is steeply inclined with a north-south strike.

Just east of Cnoc na Fuarachad, a broad valley opens out to the north, marking an outcrop of Torridonian Sandstone which ends at the road. This is succeeded farther to the east by another ridge of quartzite, which ends to the north in Sgiath-bheinn an Uird. After passing the quarry in this quartzite, the valley of the Ord River opens out to form a wide area of Torridonian Sandstone, with a much more subdued topography.

The Cambrian Quartzite continues north of the road where it is exposed in bare slopes of whitish rock, overlooking the valley to the south. The main outcrop of Torridonian Sandstone is reached beyond this point, a mile and a half from Ord. This extends east until the outcrop of the Lewisian Gneiss is encountered once more, beyond Loch Meodal. Continue east as far as the A851, turn left and return to Broadford.

BROADFORD TO ELGOL

Starting from Broadford, take the A881 towards Elgol. This road first crosses Jurassic and Triassic rocks, which are not well-exposed, and then a narrow outcrop of Torridonian Sandstone, before it reaches good exposures of Durness Limestone along Strath Suardal. There are good views of the Eastern Red Hills to the north, consisting of Beinn na Caillich, Beinn Dearg Mhor and Beinn Dearg Bheag. All these hills are red granite, which forms a single intrusion of Tertiary age. Their rounded summits fall away in steep slopes, covered with screes.

The contact of this granite with its country-rocks to the south passes across the slopes of Beinn Dearg Bheag, above Loch Cill Chriosd. There is an abrupt change in topography as the granite screes give way downhill to much rougher ground around Coire Forsaidh. The rocks exposed on these craggy slopes are mostly agglomerates, lying within a volcanic vent.

Southeast of Strath Suardal, the Durness Limestone outcrops in typical fashion around the summit of Ben Suardal. However, the northern slopes of this hill are Torridonian Sandstone, folded around an anticline with a core formed by the Durness Limestone. Exactly the opposite of what might be expected, it is evident that Caledonian earth-movements have intervened, thrusting Torridonian Sandstone over Durness Limestone before these rocks were folded together to form the anticline itself.

Farther southwest, the Beinn an Dubhaih Granite appears in the core of this fold, intruding the Durness Limestone. It makes the rather lower ground, covered in peat and heather, overlooking the road towards Torrin. Its contact with the Durness Limestone will be examined near Loch Cill Chriosd on the return journey.

Just before reaching Kilbride, make a detour along a track which leads south to *Camas Malag* on the shores of Loch Slapin. Before reaching the coast, a quarry is seen to the west, working the Durness Limestone where it has been converted into a white marble. It is cut by several dykes of dark basic rock, which are deformed to such an extent that they have broken up into separate segments.

GEOLOGICAL LOCALITY: CAMAS MALAG

On reaching the coast at Camas Malag, good exposures of the Beinn an

Dubhaich Granite can be examined along the shore. Then walking south, the granite contact can be located within a few feet where it descends a prominent gully below the track, just before the headland at [NG 583188]. The rocks in contact with the granite are marbles, representing the Durness Limestone where it has been metamorphosed by this igneous intrusion. A feature of particular interest at this point is the presence within these marbles of dark basic rocks, forming disrupted masses, that were once basalt dykes of Tertiary age. They have been deformed as a result of the earth-movements accompanying the intrusion of the Beinn an Dubhaich Granite. They are best seen on the south face of the headland, where great care is required as the rocks are very steep.

Farther south along this coast, the Durness Limestone is overlain unconformably by Jurassic rocks, which make the heather-covered slopes dipping towards the sea around the mouth of Allt nan Leac.

GEOLOGICAL PANORAMA: LOCH SLAPIN

There is an excellent view westwards across Loch Slapin from Camas Malag, with the serrated ridge of Bla Bheinn on the distant skyline. The lower ground on the far shore of this loch is occupied by Jurassic rocks, mostly sandstones and shales, which make excellent features on the terraced hillsides, descending at a low angle towards the north. They are overlain by Tertiary lavas, nearly all basalts, exposed in the steep cliffs capping the escarpments around Ben Meabost and Am Camach.

Bla Bheinn appears to the north of this escarpment. It consists of an intrusive mass of gabbro, cut by many Tertiary dykes. It is the weathering out of these dykes that gives this mountain and the other Cuillin ridges their jagged outlines, so unlike the rounded summits of the Red Hills. Such a marked difference in topography arises because only a few dykes of Tertiary age cut the later granites of the Red Hills, whereas they are much more abundant in the Black Cuillins.

North of Bla Bheinn, the gabbro continues to outcrop in the summits of Sgurr nan Each and Garbh-bheinn, while the next mountain to the north is Belig, which consists of Tertiary lavas, dipping towards the east. The high ground falls away beyond this point, where granite makes the rounded hills of Glas Bheinn Mhor and Beinn na Cro. This panorama ends east of Loch Slapin in the granite hills of Beinn Dearg Mhor and Beinn Dearg Bheag.

Returning to the A881, continue towards Elgol around the head of Loch Slapin. The road crosses Jurassic rocks as it runs along the eastern side of the Strathaird peninsula. These rocks are quite well-exposed along the road, giving rise to excellent dip-and-scarp features in several places. At Kirkibost, there is a good view westwards to the higher ground formed by the Tertiary lavas, showing the layered nature of this volcanic sequence. On reaching Elgol, stop in the car-park [NG 519137] and walk down the road to the pier.

GEOLOGICAL EXCURSION: LOCH CORUISK

During the summer months, a geological excursion can be undertaken by motor-boat from Elgol to Loch Scavaig in the heart of the Black Cuillins. The headlands guarding the entrance to this sea-loch are Torridonian Sandstone, overlain by Tertiary lavas. Beyond are found the igneous rocks which make up the intrusive complex of the Black Cuillins. These rocks consist mostly of coarse-grained gabbros, together with some ultrabasic rocks known as peridotites, rich in olivine. All these rocks are very resistant to weathering and erosion, so that they now stand high above their surroundings, forming the magnificent ridge of the Black Cuillins with its 24 peaks, each over 3000 feet in height.

Once the Black Cuillins must have been much higher, but glacial activity has been intense, carving out great corries and deep rock-basins, and leaving the ridge itself as a narrow and often sensational arete, running from peak to peak. The effects of such glacial erosion are very obvious on landing at Loch Scavaig, where the rock-bar separating Loch Coruisk from the sea has been scoured by the ice, leaving a very steep slope on its seaward side. These exposures show gabbro, which is often layered owing to varying amounts of feldspar and pyroxene in the rock. Farther afield, peridotites are exposed around An Garbh-coire, weathering to a distinctive orangey-brown.

Returning by boat from Loch Scavaig, the island of Rhum can be seen to the south, together with the Island of Canna to its west. Rhum is another Tertiary complex like the Cuillins, while Canna is composed of Tertiary lavas. After landing at Elgol, where an excellent example of honeycomb weathering (**please do not hammer**) can be examined on the headland just north of the school, return along the A881 towards Broadford. After passing through Torrin, park at the roadside at the eastern end of Loch Cill Chriosd.

GEOLOGICAL LOCALITY: LOCH CILL CHRIOSD

This short excursion starts where a rough path leaves the road, 400 yards west of the ruined church at *Kilchrist*. Walk 100 yards west to where a Tertiary dyke can be examined at the roadside [NG 614203]. Up to 100 feet in width, this vertical mass of dark igneous rock intrudes the Durness Limestone. It makes a ridge, running southeast away from the road, covered in heather. The lower ground on either side of this ridge is limestone, giving rise to grassy slopes.

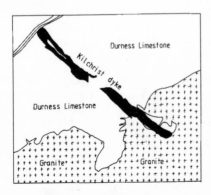

Figure II. Geology around Loch Cill Chriosd.

Climbing up from the road, a raft of limestone can be found within the dyke where a power line crosses its outcrop. Note how this rock weathers in a manner typical of limestone almost everywhere. The dyke itself can be traced as a continuous feature for 250 yards towards the southeast, until it suddenly stops. There is a break of 50 yards in its outcrop at this point, marked by exposures of limestone, beyond which it resumes its southeasterly course. The dyke itself continues for a short distance towards the east as a narrow intrusion of igneous rock, ending just before the fence is reached. After crossing this gap in its outcrop, the dyke can then be followed for another 300 yards towards the southeast. Limestone forms a grassy hollow to its northeast, along which there are ruins of old crofts. The dyke ends just south of the last croft.

Beyond this point, granite is encountered along the outer contact of the Beinn an Dubhaich Granite. This igneous intrusion is well-exposed to the south, giving rise to very boggy moorland covered in heather. Its contact with the limestone northeast of the dyke can be followed where heather gives way to grass. The contact between dyke and granite is less easy to establish. However, the dyke consists of dark rocks, much jointed and covered with moss and lichen, while the granite is much lighter in colour, lacks much jointing and forms rather smooth exposures. As the dyke cannot be followed into the granite, it was presumably intruded before the granite, which now cuts across this earlier intrusion. The granite itself forms a curious embayment along the line of the dyke.

Now walk northwest back along the dyke, until an old wall is reached. The granite contact follows the dyke immediately to the southwest, except that there are occasional outcrops of limestone as well. Just beyond the wall the heathery ground underlain by granite gives way to limestone, covered in grass. The contact of the granite with the limestone can then be traced towards the southwest from this point. Tongues of granite penetrate the limestone in a very irregular manner. However, this contact becomes more regular, some 80 yards away from the dyke. Its presence is emphasised by an abrupt change in the vegetation. Note how weathering and erosion has lowered the surface of the limestone by a few feet, presumably since the end of the Last Ice Age. One consequence is that a small stream flowing over the granite descends in a waterfall as it cross the contact with the limestone, only to disappear down a sink-hole into the limestone itself.

Mineralisation has occurred in several places along this contact where the granite has reacted with the limestone. The clearest example is seen just beside an abandoned mine-shaft, filled with water, where the rock is rich in magnetite, an oxide of ferric iron. Specimens of this dark and very dense rock clearly affect a compass needle. There are also traces of bright green malachite, a copper mineral. Any mineral collecting should be restricted to some spoil heaps, which are found 25 yards northwest towards the ruins of the Old Manse.

BROADFORD TO SLIGACHAN

Returning to Broadford, follow the A850 road north towards Sligachan. It first skirts the Tertiary intrusions, mostly granites of various kinds, that are so well-exposed in the Red Hills to the west. However, there is little exposure along the road before Camas na Sgianadin, where the Broadford Gabbro occurs in a cutting opposite a large layby [NG 622258]. Beyond this point, there is a thin strip of light-coloured granite, followed by altered lavas near a disused quarry at Strollamus.

The road then crosses the valley of Allt Strollamus, where there is a narrow strip of Jurassic rocks, consisting of sandstones, limestones and shales, all much altered by the nearby intrusions. Beyond this point, granite is reached just before Dunan, and the road crosses its outcrop farther around the coast towards Loch Ainort. An outlier of this granite is seen on Scalpay in contact with Torridonian Sandstone, which otherwise makes up the greater part of this island.

The road then runs inland at the head of Loch Ainort, crossing over more granite. Good exposures are seen at the roadside below Bruach nam Bo, and beyond Allt Mhic Mhoirein. Descending Gleann Torra-mhichaig towards the coast, there is much hummocky moraine, covering the lower slopes of this valley. This obscures the contact of the granites with their country-rocks, which appear as the road approaches Sconser.

The view north along the coast from this point shows trap featuring where the Tertiary lavas outcrop on the hillsides above Peinchorran. Ben Tianavaig is seen in the distance as a sharp-topped hill, consisting of Tertiary lavas dipping towards the west. Its eastern slopes are formed by a huge land-slide, formed where the Tertiary lavas have collapsed towards the sea on the underlying Jurassic rocks.

Around Sconser, the road passes exposures of well-bedded Jurassic lime-stones and shales, dipping towards the northwest, before it crosses on to Tertiary lavas. These dark-looking and very blocky rocks are intensely altered in the aureole of the Glamaig Granite to the south. They can be examined in roadside exposures at the layby, just before crossing the river near Sligachan. Park near the hotel.

GEOLOGICAL PANORAMA: SLIGACHAN

The lower ground around Sligachan [NG 48529] is blanketed with hum-mocky moraine, left by the glaciers when they melted at the end of the last glaciation. These deposits obscure the Tertiary lavas, which form the low ground at the mouth of Glen Sligachan as far south as Nead na h-Iolaire. However, they can be seen to the northeast, where they are exposed in the escarpment of An Leitir, overlooking Loch Sligachan to the northwest. These lavas give way at Nead na h-Iolaire to the intrusive rocks of the Cuillins Complex, which make the shapely peak of Sgurr nan Gillean, west of Glen Sligachan. These rocks are mostly gabbros, giving the landscape quite a som-bre appearance wherever they are exposed.

The hills east of Glen Sligachan are very different in character. They con-sist of steep-sided hills of light-coloured granite, much less well-exposed than the gabbros of the Black Cuillins, but covered with much scree. The dark-coloured gabbro of Bla Bheinn is visible in the distance at the far end of Glen Sligachan, beyond Marsco.

SLIGACHAN TO BERRERAIG BAY

Take the A850 road north from Sligachan towards Portree, crossing over Tertiary lavas for the whole distance. These lavas give rise to terraced hillsides with trap featuring, as the tops of these lava-flows often consist of rubbly rock much more suceptible to weathering and erosion than their more mas-sive centres. As the road descends Glen Varragill, the Old Man of Storr appears in the far distance, standing proud of the lava escarpment at The Storr. As it is rarely possible to get a good view of the coast along the Sound of Raasay, north of Portree, a detour can be made along the B883 to Peinmore, and then by turning left along a side road to Penifiler.

Parking near the end of the road, walk north over rough moorland to the slopes above Camas Ban. There, the view north towards Torvaig shows Tertiary lavas exposed in the headland of Ben Chracaig, and farther north in the steep cliffs that make the escarpment beyond Rubha na h-Airde Glaise. Sedimentary rocks of Jurassic age are exposed on the lower slopes below this escarpment. They are brought up to the southwest by a fault running inland from the coast, just where the escarpment ends. These rocks give the gently sloping ground that is open to cultivation around Torvaig. However, they are faulted down again just east of Ben Chracaig, where they are lost to view. Returning to the main road, drive through Portree and take the A855 north towards Staffin.

After passing through open country just north of Portree, the road comes into view of the lava escarpment which extends with hardly a break for nearly 15 miles from A'Chorra-bheinn in the south to Sgurr Mor in the north. This escarpment forms the back of a huge landslip affecting the Tertiary lavas in the north of Skye. The landslip itself makes up the lower and often very broken ground lying at the foot of this escarpment. The Tertiary lavas within this landslip have moved downhill towards the east, sliding on weak horizons in the underlying Jurassic rocks, as these rocks were eroded by the sea.

GEOLOGICAL LOCALITY: BERRERAIG BAY

Driving north past Lochan Leathan, a private road leads east to the dam at the northern end of this reservoir. Crossing this dam, continue down the road to reach the flight of 640 steps which descends by the pipeline to reach the hydro-electric power station at Bearreraig Bay [NG 517527]. The coast just round the headland to the east provides an excellent section through the Jurassic rocks lying below the Tertiary lavas of The Storr. **Beware of incoming tides, and take care below high-water mark as the rocks are very slippery with seaweed.** The cliffs along this section are formed by calcareous sandstones with large concretions of carbonate, while the foreshore exposes the underlying shales, rich in fossils. These include the coiled shells of ammonites, the narrow cones of belemnites, and ordinary bivalves.

Two igneous dykes are seen close to low-water mark, where they cut across the bedding of these sedimentary rocks. The more northerly dyke shows columnar jointing and chilled margins against its country-rocks, which have been hardened by the heat of the intrusion. The other dyke has a selvedge of darker country-rock along its contacts, where the surrounding shales are altered. Climbing back up the pipeline, the escarpment north of Berreraig Bay is seen, capped by a Tertiary sill with columnar jointing.

GEOLOGICAL LOCALITY: OLD MAN OF STORR

Returning to the main road, there is a car-park 500 yards farther north, where a path leads up through a small plantation towards the Old Man of Storr [NG 500539]. This is an isolated pinnacle of Tertiary lava, 160 feet high and 40 feet in diameter, which stands in front of the main escarpment at The Storr. It is only one of a number of rock pinnacles, isolated from one another by weathering and erosion. The escarpment itself exposes a sequence of 24 lava-flows at the very least, forming the cliff-face which reaches its highest point in The Storr. The face is also seamed by vertical gullies, weathering out along basalt dykes where they cut the volcanic sequence.

The lava-flows are mostly basalt, rich in gas-bubbles. These cavities are now filled with zeolites, which are pale-coloured minerals forming hydrous alumino-silicates of sodium, potassium, calcium, barium and strontium. Good specimens can be collected from the scree slopes just beside the path, southwest of the Old Man of Storr. Blocks of bright red laterite are also present in these screes, representing the weathered tops of lava-flows.

Continue up the path from the Old Man of Storr, passing to the left of another pinnacle, known as the Needle, to reach its highest point, just south of a fence. The individual masses of Tertiary lava, forming the landslipped area around the Old Man of Storr, can best be appreciated by looking south from this point. To ascend The Storr, follow the path as it curves northwest into the shallow corrie northeast of the summit, which can then be reached by climbing uphill towards the southwest. **Great care should be taken as the cliff-edge is unstable.** Return back along the same route to the road.

NORTH OF TROTTERNISH

Continue north along the A855, across rolling moorland with vertical cliffs overlooking the Sound of Raasay to the east. There is a good view towards Raasay where the road crosses the steep-sided gorge of the Lealt River, showing how this river has cut down into the Tertiary sills that form the high cliffs along this part of the coast. Farther north, the dolerite forming these Tertiary sills is exposed in roadside cuttings, just before Culnaknock is reached. Beyond this point, the Iron-Age fortifications of Dun Dearg and Dun Grianan are built on abrupt hills of dolerite, forming outliers of a Tertiary sill lying on top of Jurassic rocks. Stop in the car-park just north of Dun Grianan at the mouth of Loch Mealt.

This viewpoint [NG 509655] looks north to the *Kilt Rock*, so-called because the Tertiary sill making the cliffs at this point displays columnar jointing, looking very like the pleats in a kilt. The sill itself is composed of dolerite, forming very dark rocks in marked contrast to the Jurassic rocks which it intrudes. These sedimentary rocks are much lighter in colour, while their bedded nature is quite clear, even at a distance.

The view to the south shows the sill which forms Dun Grianan and Dun Dearg at the very edge of the cliff, while another sill is present lower down the cliff. This sill can be seen through binoculars to change its level, evidently cutting across the bedding of the Jurassic rocks as an intrusion, rather than a lava-flow. By walking back along the main road, the dolerite forming these intrusions can be examined at close quarters.

Continue north from Loch Mealt to Staffin, and then turn left along the minor road to The Quiraing. This road ascends the escarpment above Loch Leum na Luirginn in a series of hairpin bends. Park where the road reaches the top of the pass [NG 440679]. Walk back east down the road to a small quarry, where the reddened top of a typical lava-flow is exposed. Farther down the road, a path leads north along a terrace to *The Quiraing*.

All this ground is a huge landslip, forming great wedges of Tertiary lava, each resting on a curved surface, which flattens out towards the lower ground farther east. As these wedge-like masses of Tertiary lava slipped down these surfaces, their upper surfaces became tilted back in the opposite direction towards the west in a chaotic fashion. Pinnacles are formed where these tilted lava-flows have been attacked by weathering and erosion.

Return down the road to Staffin, and turn north along the A855 towards Flodigarry, passing across the foot of the landslip. After nearly two miles, the roadside exposures around *Dunans* [NG 467706] show a sequence of several lava-flows, each with a reddened top (MFG 80), dipping towards the south-east. Just beyond this point, close to Loch Langaig, there is a good view towards the southwest, showing the landscape created by this landslip.

Continue along the road towards Uig around the northern tip of Trotternish. The topographic features seen throughout this area are mostly formed by Tertiary sills, standing out as much more resistant masses in comparison with the intervening Jurassic sediments. This can be seen at *Duntulm Castle* [NG 410744] which stands on dolerite, while the bay just to the south is underlain by Jurassic rocks.

Continuing south, the cliffs overlooking the road along the eastern shore of Lub Score expose Tertiary dolerites with columnar jointing, from which huge blocks have fallen. Tertiary lavas then appear once more around *Totscore*, approximately 5 miles to the south, where they are seen in a quarry just east of the road at [NG 388661]. This is another locality showing the reddened tops of lava-flows, formed as usual by intense weathering under subaerial conditions, most likely in a tropical climate.

After another mile, the road descends to the village of Uig, where Jurassic rocks are found around the bay, faulted against the Tertiary lavas to the north. The road then climbs the escarpment to the south, giving good views of the Tertiary lavas where they are well-exposed in the headlands of Ru Idrigill and Ru Chorachan, guarding the entrance to Uig Bay. The layered nature of these volcanic rocks is clearly visible, particularly on the cliffs to the south.

Follow the A856 towards Portree, crossing Tertiary lavas for the whole distance. The terraced nature of the hillsides, and the occasional exposure showing the reddened top of a lava-flow are characteristic features of this ground. On joining the A850, turn right at Portree and continue through Sligachan to Kyleakin and Broadford to reach Kyle of Lochalsh. Alternatively, the ferry to Glenelg can be taken from Kylerhea during the summer, continuing along the trail from Kyle of Lochalsh at Shiel Bridge, or the crossing can be made by ferry to Mallaig from Armadale, rejoining the trail from Fort William at Lochailort.

CHAPTER 10

Kyle of Lochalsh to Fort William

THIS SECTION OF THE Highland Geology Trail passes from the Moine Thrust Belt around Kyle of Lochalsh eastwards to cross the Moine Thrust. Beyond the Moine Thrust are exposed the Moine Schists, forming a wide outcrop of metamorphic rocks, cut by granite intrusions. The trail then follows the Great Glen Fault to Fort William, apart from a detour along Glen Spean to view the Parallel Roads of Glen Roy. It then turns west, crossing the Moine Schists once again to reach the Tertiary Igneous Centre of Ardnamurchan.

KYLE OF LOCHALSH TO FORT WILLIAM

Follow the A87 east from Kyle of Lochalsh towards Shiel Bridge. The route first crosses Torridonian Sandstone, overturned towards the west above the Kishorn Thrust. The outcrops along the road show greenish-grey rocks, dipping steeply towards the east, cut by many quartz-veins. These rocks have lost their reddish colour owing to the low-grade metamorphism that accompanied the Caledonian earth-movements. Farther east towards Balmacara, these rocks become very flaggy and quite flat-lying, as the Moine Thrust is approached, carrying Lewisian Gneiss westwards over the Torridonian Sandstone.

The overthrust rocks are seen farther east along the road beyond Auchtertyre, where banded gneisses are exposed in a series of roadcuts. Passing through *Dornie*, these gneisses are then very well-exposed along the A87 for nearly three miles as far as the parking place at An Leth-allt. **If stopping, care should be taken on account of fast traffic and loose rocks.** More details are given in the Geologists' Association *Guide to the Northwest Highlands*. The main outcrop of the Moine Schists is reached around Inverinate, but these rocks are not well-exposed along the road. Continuing towards Shiel Bridge, an outlier of the Ratagan Granite is crossed, giving roadside exposures of very pink granite near Kintail Lodge.

Southeast beyond Shiel Bridge, the A87 ascends *Glen Shiel*, where the Moine Schists are well-exposed on the steep hillsides. These rocks are quite flaggy with steep dips towards the northwest in most places. The complex nature of the folding and deformation affecting these rocks can be seen by stopping at a large layby on the south side of the road, 8 miles beyond Shiel Bridge at [NH 027123], and walking uphill over rough ground to the southwest.

Two quite substantial streams need to be waded to reach these exposures. The second stream is the *Allt a'Choire Reidh*, which is best crossed 100 yards upstream from the bend where it changes course towards the north. Walk uphill towards the west from this point, keeping to the right of the crags that can be seen from the road, and then climb the slope to reach the top of these crags around [NH 020120]. The ridge at this point is crossed by a gully, and the best exposures lie just uphill to the southwest.

Continue along the A87 to the head of Glen Shiel, where more open country is reached, east of the watershed. The Moine Schists exposed along the road to the north of *Loch Cluanie* show migmatites [NH 103115] and granite veining [NH 120106], before reaching their contact with the Cluanie Granite, 4 miles beyond Cluanie Inn. There is an abrupt change in the landscape at this point, grassy slopes giving way to much more rock, eroded into smooth outcrops as the ice passed over this granite. Beinn Loinne to the south, and Carn Ghluasaid to the north, are both hills carved in typical fashion out of the Cluanie Granite. The road then crosses back on to Moine country-rocks, just after the hydro-electric dam is passed at the eastern end of Loch Cluanie.

Follow the A87 where it turns right towards Invergarry at its junction with the A887, and cross the watershed into Glen Garry. Roadcuts show that the Moine Schists are flat-lying throughout much of this area. A detour can be made to Loch Quoich by turning right where the A87 road reaches the shores of Loch Garry, along the minor road to Tomdoun and Kinloch Hourn. After reaching the dam at the foot of Loch Quoich, continue for 10 miles to a parking place at a bend, just below a radio mast at [NH 044018].

Walk downhill to the southeast from this point, heading for the end of a promontory sticking out into Loch Quoich. The shore-section along the southwest side of this promontory shows several generations of folds, together with complex patterns of boudinage, all magnificently exposed. **These exposures make a sloping surface polished smooth by the ice, so take care, especially in wet weather.** This locality is best visited when the level of Loch Quoich is low, although this is not essential.

Return to the A87, and continue east to Invergarry. Turn right and follow the A82 along the Great Glen towards Fort William. This road follows the deep valley eroded by the ice along the line of the Great Glen Fault. Turn left at Spean Bridge along the A86 towards Roy Bridge, where another left turn should be made along the road leading to *Glen Roy*. After 3 miles, stop in the car-park at the viewpoint [NN 297853].

GEOLOGICAL PANORAMA: GLEN ROY

This well-known viewpoint looks north to the Parallel Roads of Glen Roy. These are the ancient shore-lines of a lake, dammed by glaciers towards the end of the Last Glaciation. Although the ice had mostly melted away by 12,000 BC, the climate became much colder around 9,000 BC, when glaciers once more became established, although not so widely as before. Ice from beyond the Great Glen flowed east into the valley of the River Spean, damming the mouths of the glens to the north. This impounded the waters of a lake, filling Glen Roy and its nearby valleys to a depth of several hundred feet.

The highest shore-line lies at 1,150 FEET OD (or 350 metres), corresponding in height to the col at the very head of Glen Roy. The lake then drained eastwards into the head-waters of the River Spey. However, as the glaciers started to retreat around 8,000 BC, they uncovered another col at

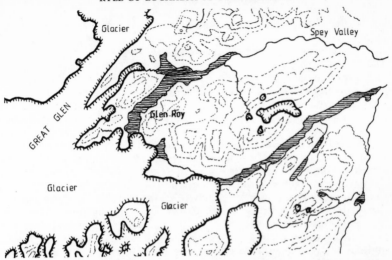

Figure 12. The Parallel Roads of Glen Roy, showing the Marginal Lakes dammed by the Glaciers at the End of the last Ice Age.

1,075 feet OD (or 325 metres). This lies to the east of Glen Roy at the head of Gleann Glas Dhoire, and the lake drained east into the valley of Feith Shiol. The middle shore-line in Glen Roy corresponds to the height of this col. Finally, the ice retreated so that the lake was able to drain directly into the Spey valley across the col at the head of Glen Spean. The height of this col is 850 feet OD (or 260 metres), corresponding to the height of the lowest shore-line in Glen Roy.

After exploring Glen Roy, return to Spean Bridge, and continue along the A82 towards Fort William. Just before entering this town, turn west along the A830 towards Mallaig.

FORT WILLIAM TO ARDNAMURCHAN

The A82 road from Fort William crosses the Moine Schists lying to the west beyond the Great Glen Fault, although these rocks are poorly exposed. Beyond the head of Loch Eil, the geology becomes more complex towards Glenfinnan, where much granitic material is present. These rocks are best examined by parking at a layby [NM 916802], opposite a rocky bluff, covered in pine-trees. The roadcut at this point exposes granite gneiss, with thin quartzo-felspathic veins separated from one another by more micaceous rock. There are also very conspicuous veins of light-coloured pegmatite exposed all along this road.

Continuing past Glenfinnan, make another stop 3 miles to the west, where the road starts to descend sharply towards Loch Eilt near *Creag Ghobar*. Park in a layby at [NM 857816] on the south side of the road. Exposures on the hillside north of the road show the complex folding which affects the Moine Schists, together with abundant veins of pegmatite. They can best reached by walking a short distance back up the road, and then ascending a slight valley just to the left of a prominent boss of rock. This leads to a sloping shelf of rock, forming a very extensive glacial pavement. The ground drops away to the west in a steep rock-face, plucked by the ice as it flowed in the same direction across this watershed. The best exposures lie just above this rock-face, overlooking the glacial trough of Loch Eilt.

Return to the road, and continue along the A830 past Loch Eilt. Park in a layby a few hundred yards beyond its junction with the A861 at Lochailort.

GEOLOGICAL LOCALITY: LOCHAILORT

The roadcuts along the new section of the A830 just west of Lochailort expose Moine rocks that were once impure but finely-bedded sandstones. They dip steeply towards the northwest. The second roadcut [NM 762824] from the junction with the A861 shows that the rocks on the north side of the road are cut by thin sheets of darker rock, dipping southeast at 30 degrees. By matching up the country-rocks across these intrusions, it can be seen that the overlying rocks have moved up-dip to the northwest, while the underlying rocks have shifted down-dip to the southeast. The presence of a schistosity within these thin sheets of igneous rock suggests that these movements must have occurred after their intrusion.

Return to the road-junction at Lochailort and take the A861 towards Kinlochmoidart and Acharacle. The road crosses Moine Schists for virtually the whole way to Salen, cut by many Tertiary dykes. Turn right at Salen along the B8007 towards Kilchoan. Tertiary dykes also cut the Moine Schists at intervals all along this road. They are clearly seen at *Rubh' an t-Sionnach* [NM 662612], appearing as rather massive rocks with horizontal jointing, weathering to a dark brown.

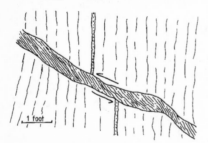

Figure 13. Schistose Dykes at Lochailort.

Once past Glenborrodale, there are views across Glenmore Bay to the Tertiary lavas capping the summit of Beinn Bhuidhe. After another two miles, the road crosses on to these lavas, and the landscape changes abruptly. These basalt lavas weather easily to a rich brown soil, giving much more grassy slopes than the Moine Schists. Where the road turns inland, there is a good view west to Ben Hiant and Maclean's Nose. This hill is the site of a volcanic vent, lying just east of the Tertiary igneous complex, centred on Ardnamurchan itself. The ground is poorly exposed as the road passes to the north of Ben Hiant, eventually reaching the village of Kilchoan.

The Tertiary complex of Adrnamurchan consists of several intrusions, mostly of gabbroic rocks, which form a series of concentric outcrops around a volcanic centre. These intrusions affect the topography in different ways, but there is a particularly wide ring-dyke, known as the Great Eucrite, which makes a ring of very prominent hills in the centre of the peninsula. This can best be appreciated by taking the road northwest from Kilchoan, which turns right towards Sanna after little less than a mile.

Two viewpoints can be recommended. The first is *Creag an Airgid*, lying just east of the road, 2 miles north of Kilchoan. The second is reached by walking half a mile east from *Achnaha* towards rising ground, where the very centre of this igneous complex is marked by a small cairn. The Great Eucrite forms a vast amphitheatre around this point, overlooking the lower ground formed by the less resistant rocks in its centre.

Continue along the road towards *Sanna*, where the effects of glacial erosion are particularly clear around [NM 455690]. Similar features are seen on the slopes of Creag an Airgid. Further exploration is beyond the scope of this trail, but the visitor is referred to the Ardnamurchan guide, published by the Edinburgh Geological Society.

ARDNAMURCHAN TO FORT WILLIAM

Return along the same road to Salen, and take the A861 towards Strontian and Corran Ferry, passing across the Moine Schists for the first 5 miles. Tertiary dykes are seen on the shore, just after Resipole. The contact of the Strontian Granite is then crossed near Sron na Saobhaidh, just over 2 miles before Strontian. Glaciated exposures of this granite, which carries flattened fragments of its country-rocks, are found along the shore at **Ranachan** [NM 790611], and farther east towards the head of Loch Sunart.

The road continues through Glen Tarbert, which has a U-shaped profile, typical of a glacial trough. The floor of this valley is covered with hummocky moraine, while it is crossed by a terminal moraine which forms a prominent ridge of gravelly material, 2½ miles east of the road-junction with the A884. The flat ground to its east was once the floor of a lake, dammed by this moraine. There is an alluvial fan at the watershed to the east, consisting of quite large boulders, which were dumped by the stream draining from the north.

At the eastern end of Glen Tarbert, where it reaches the shore of Loch Linnhe, the road turns northeast along the line of the Great Glen Fault. The shattered nature of the rocks affected by this fault can be seen in virtually every exposure along the road. The best view of the deep valley that forms the Great Glen to the northeast is seen just north of *Corran Ferry*.

Raised beaches and fluvio-glacial terraces are also a common feature along this coast. In particular, there are wide terraces of gravel on both sides of Loch Linnhe at Corran Ferry, standing around 75 feet above sea-level. Several deep lochans occur behind Corran where large masses of 'dead ice' melted away, leaving what are known as kettle-holes. Crossing Loch Linnhe at Corran Ferry, turn left along the A82 into Fort William.

CHAPTER 11

Fort William to Oban

THE TRAIL NOW ENTERS THE Southwest Highlands, southeast of the Great Glen Fault. The nature of the geology is now quite different. The Southwest Highlands consist mostly of Dalradian rocks, folded and metamorphosed in a complex fashion around 600 million years ago, to judge by the latest research. They were originally a very varied sequence of sedimentary rocks, that are now altered to quartzites, metamorphic limestones or marbles, and slates, phyllites or schists, together with basic igneous rocks, which now occur as high-grade amphibolites or low-grade epidiorites.

All these Dalradian rocks were intruded by granites towards the start of Devonian times, around 400 million years ago. These great masses of igneous rock mostly form high ground, apart from the Moor of Rannoch Granite. The intrusion of granite at depth was accompanied by volcanic eruptions at the surface, and thick sequences of lava-flows and pyroclastic rocks accumulated on top of an eroded surface of Dalradian rocks. These Devonian rocks are now preserved in the cauldron-subsidences of Ben Nevis and Glencoe, and farther south as the Lorne Lavas, resting unconformably on top of Dalradian rocks.

The lower slopes of Ben Nevis in particular are formed by granite of Devonian age, which intrudes the surrounding rocks. These country-rocks are Dalradian limestones and schists, now metamorphosed by the granite into a hard, splintery rock known as a hornfels. The granite itself occurs as two separate intrusions. The outer granite is a coarse-grained pinkish rock, containing large crystals of felspar. This granite is cut by a dyke swarm, centred on Ben Nevis.

These dykes, however, do not cut across the inner granite, so that this intrusion must be later than the outer granite. The inner granite lacks the large feldspars found in the outer granite, so that it is a finer-grained and more even-textured rock. The upper slopes of Ben Nevis are andesite lavas and agglomerates, around 2,000 feet in thickness. These Devonian rocks form a plug that sank into the inner granite, probably as it was intruded upwards in a molten state.

GEOLOGICAL EXCURSION: BEN NEVIS

The tourist path up Ben Nevis starts from a foot-bridge which crosses the River Nevis opposite Achintee House. This can be reached from Fort Willian by turning right off the A82, just north of the town, and driving along Glen Nevis for a mile to a car-park [NN 123730]. The weather on Ben Nevis can be treacherous, and any party should be properly equipped.

After passing Achintee House, the tourist path first crosses the meta-morphic rocks lying in the aureole of the granite. These rocks are not well-exposed, but they can be seen in small outcrops opposite the camp-site [NN 129724]. They are pale-green rocks, rich in what are known as calc-silicate minerals, formed by the alteration of impure limestones. After crossing a stile, exposures of the outer granite are next seen along the path, cut by dykes of dark rock, belonging to the Ben Nevis swarm.

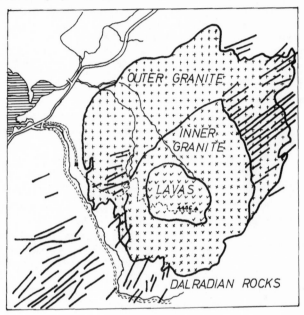

Figure 14. Geological Map of Ben Nevis.

The granite also contains patches of much darker rock, which are altered fragments of country-rock, known as xenoliths, torn from the walls of the intrusion. There are good views up Glen Nevis beyond this point. In particular, the reddish screes on Mullach nan Coirean, another granite intrusion, make a strong contrast with the grey screes of Dalradian quartzite on Sgurr a'Mhaim.

The contact between the outer and inner granites is crossed and recrossed as the path zig-zags up the slopes south of Lochan Meall an t-Suidhe. On reaching the slopes above this lochan, the path then remains on the inner granite, which is best exposed where the path turns back towards the south. After crossing the Red Burn, the path zig-zags up the steep slopes below the summit. The volcanic rocks are first seen after the third zig-zag around a height of 3,000 feet (915 metres). They continue all the way to the summit plateau, which is covered with angular fragments of andesite lava, shattered by frost.

Return down the tourist path to the car-park, and drive up Glen Nevis to *Polldubh* where the road crosses the river. Two dykes of dark igneous rock intrude the Mullach nan Coirean Granite just below this bridge, but otherwise the bed of the river is formed by granite. As they are less resistant to erosion than the granite, the river has cut its channel along these dykes, making two waterfalls where the river pours over the granite.

The crags northeast of Polldubh are another type of hornfels, formed by the thermal alteration of slaty country-rocks by the surrounding granites. They are very hard rocks, which the ice has eroded into roches mountonees with exceptionally smooth outlines. An excellent example is seen just south of the road, nearly a mile farther on, surrounded by pine trees. On reaching the upper car-park, there is a fine walk up the gorge of the River Nevis to the waterfall of An Steall.

BEN NEVIS TO GLENCOE

Return to Fort William, and take the A82 south towards North Ballachulish. The road first follows the shore of Loch Linnhe, itself excavated by the ice along the line of the Great Glen Fault. However, just before Onich, the road turns inland towards Loch Leven, where the Dalradian Schists are well-exposed.

The quartzites lying within this sequence of metamorphic rocks make prominent features in the landscape, particularly towards the east, where mountains of Dalradian quartzite rise steeply above the inner reaches of Loch Leven. The limestones, slates or schists elsewhere within the Dalradian sequence give a much more subdued topography, especially around Ballachulish, as they are often poorly exposed. These metamorphic rocks are intruded by the Ballachulish Granite, which forms the imposing ridge of Beinn a Bheithir, south of Loch Leven.

After crossing the bridge at Ballachulish, turn left at the roundabout along the A828 towards Oban. Stop where convenient after 2 miles, and descend to the shore around *Rubh' a'Bhaid Bheithe* [NN 024595], where exposures of the Ballachulish Granite can be seen. Grey in colour, it is a coarse-grained igneous rock, which carries fragments of metamorphosed country-rock known as xenoliths (MFG 95). The granite is cut by closely-spaced but irregular sheet-joints, all dipping towards the sea at much the same angle as the slope of the land (MFG 99).

Return to the road, and drive southwest for 1½ miles to *Kentallen*, where cars can be parked on the old road, just before the first house [NN 010578]. The roadcut exposes a coarse-grained and very dark igneous rock known as kentallenite, rich in minerals like olivine, augite and biotite, together with some feldspar. By walking back along the road and descending to the shore where the exposures start, the contact of the kentallenite with its country-rocks of Dalradian quartzite can be located around [NN 011582]. This locality also shows two igneous intrusions, cutting across one another (MFG 90).

Along the shore to the northeast, the quartzite gives way to a typical hornfels, formed by the thermal metamorphism of a slate in the aureole of the Ballachulish Granite. It is a hard, purplish rock of rather massive texture, quite unlike its parent.

Return back along the A828 to rejoin the A82, and continue east along this road towards Ballachulish. Just beyond the village are slate quarries where the Ballachulish Slate was once worked. These black slates are followed in their turn by the Ballachulish Limestone, which is exposed in road-cuts along the new road just east of the village. The view towards the east is dominated by the Pap of Glencoe, which is Dalradian quartzite.

Follow the A82 towards Glencoe, where the road turns inland. The slopes of Sgorr nam Fiannaidh to the northeast are all Glencoe Quartzite, while the floor of the glen, and the slopes of Meall Mor to the southwest, are mostly Ballachulish Limestone. However, once the road turns east opposite Signal Rock, it enters into the heart of what is known as the Glencoe Cauldron-Subsidence. After passing the National Trust Information Centre, turn left at the bridge where the road crosses the River Coe, and park immediately on the right at a small quarry [NN 137567].

Glencoe resembles Ben Nevis in its geological history, except that it is even more complex. The rocks forming the mountainous district around Glencoe are nearly all lava-flows and pyroclastic rocks of Devonian age, lying within an oval area, just 9 miles by 5 miles in extent. They are surrounded on all sides by a ring-fault, except where the Etive Granite has invaded the volcanic complex to the southeast. The volcanic rocks lying within this ring-fault foundered by several thousand feet into an underlying magma chamber, as its roof gave way, even while the volcanic activity continued at the surface.

They are now surrounded by a discontinuous ring of granitic rock, which forced its way up the ring-fault, displaced by the volcanic rocks in much the same way as a stopper forces water out of a bottle. This granite forms the Glencoe Fault-Intrusion, and its contact with the volcanic rocks lying within the cauldron-subsidence corresponds to the line of the ring-fault.

GEOLOGICAL PANORAMA: GLENCOE

The bridge over the River Coe provides an excellent view-point, allowing many features of the local geology to be appreciated. The parking place lies almost exactly on the line of the ring-fault, and the quarry exposes the fault-intrusion, charged with angular fragments of country-rock, mostly quartzite. The ring-fault follows the prominent gully which can be traced uphill on the south side of the glen towards the top of An t'Sron. The granitic rocks of the fault-intrusion outcrop to the right of this gully. They occur in contact with their Dalradian country-rocks farther to the west.

The rocks lying to the left of the gully are down-faulted within the cauldron-subsidence. The lower slopes mark the outcrop of the Dalradian rocks underlying the volcanic sequence itself. They are phyllites, exposed in some crags above the road, but otherwise forming rather gentle slopes, covered in grass. Uphill, the slopes become much steeper, and more rock is exposed, where these phyllites give way to the volcanic rocks of the cauldron-subsidence.

These volcanic rocks are best exposed on the west face of Aonach Dubh (MFG 70), above Loch Achtriochtan. The lowermost part of the Devonian sequence consists of andesitic lava-flows, at least 17 in number, which outcrop over a vertical distance of 1,500 feet on the precipitous lower slopes of

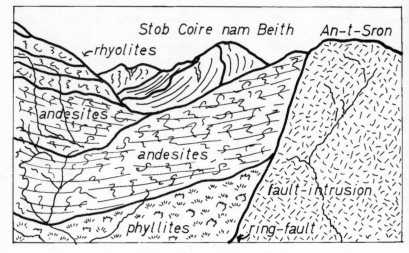

Figure 15. The Glencoe Cauldron-Subsidence.

this mountain. The succession of individual lava-flows within this sequence is clearly reflected in the terraced nature of the hillside.

Overlying these rather dark rocks are much paler rocks which form the massive crags, capping the top of Aonach Dubh. They are formed by rhyolite, which occurs as three distinct horizons, each 150 feet thick. The two lower horizons are lava-flows, while the uppermost rhyolite is a pyroclastic rock, in which volcanic glass erupted at a very high temperature has welded together the other fragmental material. All these rhyolites are very resistant to erosion so that they form the prominent and very precipitous spurs of Aonach Dubh, Gearr Aonach and Beinn Fhada, known as the Three Sisters of Glencoe.

Traced away from Aonach Dubh, these volcanic rocks come to dip more steeply around the margins of the cauldron-subsidence. This can be seen by looking south towards Stob Coire nam Beith, where pale-coloured rhyolite dips off much darker andesite at its back, away from the ring-fault where it runs behind the summit of this mountain. Likewise, the andesite lavas forming the jagged ridge of Aonach Eagach along the north side of Glencoe dip steeply towards the south, away from the ring-fault at its back.

Continue east from this viewpoint along the A82 through the heart of Glencoe. The lowermost slopes of Aonach Eagach are Dalradian phyllites, lying unconformably below the andesite lavas which outcrop in the steep and very rocky hillsides to the north of Glencoe. Ossian's Cave is seen high up on the face of Aonach Dubh to the south, where blocks of an igneous dyke have fallen away to leave a deep cleft. Farther on, andesite lavas are exposed in several roadcuts beyond Achtriochtan Farm.

Passing the Meeting of the Three Waters, the so-called 'Lost Valley' can be seen to the south. Lying between Gearr Aonach and Beinn Fhada, the valley of the Allt Choire Gabhail has been blocked by a huge rock-fall from Gearr Aonach to its northwest. This valley, like several others to the south of Glencoe, follows the same southwesterly trend as the dykes which cut the volcanic rocks of the cauldron-subsidence. A stop can be made at The Study [NN 183562], where the rhyolite lavas show good examples of flow-folding, best seen in the crags along the old road, uphill to the north. Their contact with the underlying andesites is seen at the roadside just before the Study is reached.

Continuing towards the east, Stob Gabar at the northeastern end of Buchaille Etive Bheag is mostly rhyolite. Beyond this point, the view to the south is dominated by Buchaille Etive Mor. The precipitous face of Stob Dearg, its most northeasterly peak, falls away from the summit in a series of vertical walls and great buttresses, seamed with deep gullies. The rock is rhyolite, and the gullies follow dykes of more basic rock, cutting through their country-rocks.

Stob Dearg overlooks much lower ground to its north, formed by metamorphic rocks which lie within the ring-fault. The line of this fault is crossed about a mile beyond Altnafeadh. Looking back towards the cauldron-subsidence from this point, the ring-fault can be seen where crosses the skyline at Stob Mhic Mhartuin to the north of Glencoe.

GLENCOE TO OBAN

East of Kingshouse, Glencoe opens out into the Moor of Rannoch, which is underlain by granite. Volcanic rocks are still found beyond the head of Glen Etive, where they are exposed around Sron na Creise and Meall a'Bhuiridh, south of the road. Following the slightly higher ground around the southwestern edge of the Moor of Rannoch Granite, the road passes over the contact with its country-rocks, south of Loch Ba. The low ground to the east is covered with thick deposits of hummocky moraine and many lochans.

West of the road over the Moor of Rannoch lies the very mountainous district known as the Blackmount, which extends without much break from Clach Leachad in the north to Ben Cruachan in the south. All these hills with only the occasional exception in the east are granite, making up the Etive Complex.

After passing Loch Tulla, where again the low ground is mantled with glacial moraine, turn right along the B8074 down Glen Orchy, just south of Bridge of Orchy. A quartz vein of exceptional size makes a very prominent feature on the skyline to the south. A stop can be made after 5 miles to view the pot-holed gorge of the River Orchy at *Eas Urchaidh* [NN 243321]. The river flows over garnet-mica-schists at this point, and thin stringers of quartzite have been disrupted by the folding (MFG 179). **Visitors are warned against venturing on these treacherous rocks, which become very slippery when wet, particularly if the river is in spate.**

Continuing down Glen Orchy, turn right towards Dalmally where the B8074 meets the A85. The road first crosses low-lying Dalradian rocks around Dalmally, but the view west is dominated by Ben Cruachan and the hills to the north. They are all granite, and the geology is very clearly expressed in the topography. The road skirts the granite contact as it follows the shore of Loch Awe towards the Pass of Brander, crossing Dalradian rocks which have been thermally altered by the granite to the north.

The Pass of Brander has been eroded along a fault which brought down the Lorne Lavas to the southwest by several thousand feet. These lavas are Devonian in age, consisting mostly of andesites or basalts. They are seen on the far shore of Loch Awe, exposed in the crags lying above the Pass of Brander. The road crosses the Pass of Brander Fault at the Bridge of Awe, and then continues over the Lorne Lavas as far as Oban.

In driving along the A85 from Taynuilt, stop at Connel to view the *Falls of Lora*. This is a tidal race which occurs as the tide flows in and out across a rock bar at the mouth of Loch Etive. It marks the lip of a rock-basin which was excavated by the ice as it moved down Loch Etive. The falls are best seen from the northern end of the bridge which carries the A828 north over Loch Etive towards Fort William. The Moss of Achnacree just beyond the bridge is a wide terrace of fluvio-glacial sands and gravels, complete with kettle-holes.

Farther west, the A85 enters Oban by descending the escarpment formed by the Lorne Lavas. These volcanic rocks are underlain by sandstones, shales and conglomerates, which also belong to the Lower Old Red Sandstone, resting unconformably on Dalradian slates. These slates are exposed at various places around the shores of Oban Bay, mostly below high-water mark.

CHAPTER 12

Excursions from Oban

THE GEOLOGY AROUND OBAN is dominated by the Lorne Lavas of Devonian age, which outcrop over a wide area between the Firth of Lorne and the shores of Loch Awe. Nearly all these lavas are andesites and basalts, although there are some more acid tuffs as well. These rocks are probably the same age as the volcanic rocks of Ben Nevis and Glencoe, erupted as the granites of the Southwest Highlands reached ever higher levels within the Earth's crust at the start of Devonian times, around 410 million years ago. The Lorne Lavas, together with some underlying sediments of the Lower Old Red Sandstone, rest unconformably on top of Dalradian rocks, which mostly occur as black slates and phyllites.

GEOLOGICAL EXCURSION: OBAN TO GANAVAN BAY

Negotiate the one-way system in Oban, turning north along the esplanade towards Dunollie and Ganavan. Park at the end of the esplanade near the War Memorial [NM 852309]. The coast north of this point is backed by cliffs, overlooking a rather narrow terrace of quite flat ground which lies some 25 feet above sea-level. As solid rocks are exposed along its seaward edge, this terrace is evidently a wave-cut platform of marine erosion, formed when the sea stood somewhat higher than now. The cliffs at its back are an ancient shore-line.

Walking north along the road, an old sea-stack of Lower Old Red Sandstone conglomerate can be seen standing in front of these cliffs, some 100 yards beyond the gatehouse opposite the War Memorial. The ruins of Dunollie Castle are seen a short distance to the north, standing on a bluff above this old shore-line. Although these cliffs are Lower Old Red Sandstone conglomerate, the shore exposes an outlier of the Lorne Lavas, faulted down from the east. These volcanic rocks are typical andesites, cut in some places by masses of fine-grained sediment. This has been washed into fissures in the tops of the lava-flows. The line of the fault can be seen at the north end of the section, where it forms a slight inlet [NM 852316].

Continuing along the road, there are several caves along the foot of the cliffs, north of Dunollie Castle. After a slight bend, and just before the first house at the back of Camas Ban, a wave-cut notch can be seen at the back of the raised beach, some 30 yards east of the road at [NM 853317]. There are more caves cut by marine erosion into the abandoned cliff-line north of Camas Ban, where they can be reached along a path at the back of the houses.

Continue along the road to Ganavan Bay. There are good outcrops of

Lower Old Red Sandstone conglomerate at the southern end of this bay, exposed on glaciated surfaces below high-water mark. Similar rocks are seen north of Ganavan Bay, exposed in the steep and quite high cliffs, which lie at the back of a narrow bench, cut by marine erosion at a height of some 25 feet along this rocky coast-line. The well-rounded boulders in these conglomerates are often very large, and consist mostly of dark-coloured andesitic or basaltic lava, together with some boulders of light-coloured quartzite and granite.

Return to Oban, and take the unclassified road south along the Sound of Kerrera to Gallanach. It follows a wave-cut platform along the coast, which is mostly underlain by Easdale Slates. Resting unconformably on top of these Dalradian slates are conglomerates of the Lower Old Red Sandstone, exposed in vertical cliffs at the back of the raised beach. These conglomerates are replaced to the southwest around Dun Uabairtich by andesite, making an igneous intrusion rather than a lava-flow. Just beyond the ferry to Kerrera, it displays columnar jointing of the most slender kind, with individual columns up to 120 feet long, curving slightly but close to the vertical. Cross to the island of Kerrera by the passenger ferry which runs all the year round from Gallanach.

GEOLOGICAL EXCURSION: ISLAND OF KERRERA

Kerrera provides a microcosm of local geology. The Easdale Slates form the foundations of the island, overlain unconformably by Lower Old Red Sandstone rocks. These mostly consist of thick and very coarse conglomerates at the base of the sequence, along with some sandstones and shales. They are overlain by basalt lavas, forming an outlier of the Lorne Lavas. Several faults trend northeast-southwest along the length of the island. Finally came the intrusion of basalt dykes, centred on the Tertiary igneous complex of Mull.

After crossing by ferry, first examine the rocks around the jetty, preferrably at low tide. Steely-black slates are exposed on the foreshore, just south of the pier, followed away from the sea by Lower Old Red Sandstone breccias. These rocks occur in faulted contact with one another. They are cut by a basalt dyke, weathering dark-brown, and trending towards the northwest like the other Tertiary dykes on the island.

Walk 100 yards south across a small bay to a glaciated outcrop of Lower Old Red Sandstone lava, which lies at high-water mark. The lava is a dark rock, full of gas-bubbles, now filled with secondary minerals. There are patches of sediment occupying fissures in the lava-flow, washed in from the surface just after its eruption. It is cut by a Tertiary dyke, only 18 inches thick, with chilled margins.

Walk up from the shore and turn left, following the track along the Sound of Kerrera towards the south of the island. The higher ground northwest of this track is underlain by a narrow strip of Lower Old Red Sandstone

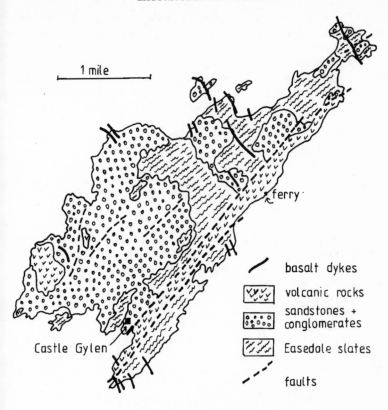

basalt dykes

volcanic rocks

sandstones + conglomerates

Easedale slates

faults

Figure 16. Geology of Kerrera.

lavas, separated by a fault from Easdale Slates to its southeast. These are exposed in the much lower ground along the shore, where the track often follows a raised beach, around 25 feet above sea-level.

After passing Ardchoirc, a Tertiary dyke makes a wall-like feature, standing proud of the Easdale Slates that are its country-rocks, 50 yards beyond the marker for a submarine cable. Beyond Little Horseshoe Bay, the track crosses the Lower Old Red Sandstone lavas as it climbs towards Upper Gylen, only to reach more Easdale Slates, just beyond the house. These slates lie northwest of the lavas, and the contact is again a fault, trending northeast-southwest.

The Easdale Slates are exposed in places along the track as it descends towards Lower Gylen, overlain unconformably by Lower Old Red Sandstone conglomerates. At the stream just before Lower Gylen, turn left along a grassy path towards Gylen Castle. On reaching Port a'Chaisteil, continue east around the raised beach to Port a'Chroinn.

The rocks on the east side of Port a'Chroinn are Lower Old Red Sandstones lavas, faulted down as a narrow strip along nearly the whole length of Kerrera. The fault separating these rocks from the conglomerates and slates to the northwest is hidden at the back of Port a'Chroinn. Just northwest of this fault is a small exposure of the Easdale Slates, overlain unconformably by the Lower Old Red Sandstone conglomerates. These conglomerates make the low headland southwest of Gylen Castle and the cliffs on which the castle itself stands.

The headland itself forms a raised beach cut by marine erosion across solid rock, while the cliffs at the back of this wave-cut platform once stood just above high-water mark. The caves at the foot of these cliffs were eroded by the sea, and there is also a natural arch, just below the castle. In front of these cliffs stand isolated stacks of conglomerate, rising above the level of the wave-cut platform. All these features date back to a time when the sea stood around 25 feet higher than it does today.

Return to Port a'Chaisteil, where the unconformity can be located near high-water mark at the back of a small bay, just below the castle ramparts. The Easdale Slates lying below this unconformity are well-exposed below high-water mark, forming the low reefs at the back of Port a'Chaisteil. They consist of dark slates with very thin beds of fine-grained, dark-grey limestone. These limestones are tightly folded in a rather irregular fashion.

These rocks are intruded by several Tertiary dykes at Port a'Chaisteil, cutting across their northeast-southwest strike almost at right angles. Some examples are very orangey-brown in colour, making a stong contrast with the rather drab limestones and slates. These dykes typically have chilled margins against their country-rocks.

Walk southwest along the far side of Port a'Chaisteil towards the very prominent stack at [NM 802264]. This stack is conglomerate, resting unconformably on the underlying Easdale Slates. There could hardly be a more perfect example of an unconformity, except that it is repeated to even better effect, some 40 yards to the west, beyond a small bay. The conglomerate is a closely-packed mass of well-rounded boulders, lying unconformably on the upturned edges of the Easdale Slates. All these slaty rocks have a cleavage, striking northeast-southwest at a high angle, along which the rock splits with relative ease.

Evidently, the earth-movements which folded and cleaved the Easdale Slates came to an end long before the overlying conglomerates were deposited. In fact, such folding and deformation must have occurred well before the prolonged uplift and erosion which so affected these rocks, that they were exposed at the earth's surface, only to be buried once more by the conglomerate and its overlying rocks.

The conglomerates lying above this unconformity form a very pronounced overhang, running along the northwest side of the small bay already mentioned. It is possible at low tide to scramble along the narrow shelf formed by the Easdale Slates where they have been eroded away by the sea below this overhang. On reaching a slight headland at the far end of this overhang, the overlying rocks are seen to be breccias, carrying angular fragments of the underlying rocks, often reddened by weathering. They rest on a very irregular surface, cutting across the bedding of the Easdale Slates, which mostly consist of limestone rather than slate at this point. These breccias pass upwards into conglomerates, if they are traced 100 yards to the northwest.

Follow these conglomerates northwest around the eastern shore of the next bay. They are cut by a Tertiary dyke, 3 feet in width, which makes a deep cleft, just south of the fence near the head of the bay. The crags stretching southwest from Lower Gylen towards the coast at this point are composed of dark sandstones, carrying much volcanic detritus, which gives them a very sombre appearance. The westerly dip shown by these sandstones brings them down to sea-level at the back of a small bay at [NM 798267], where they overlie the conglomerates at the base of the Lower Old Red Sandstone sequence. The dark-grey sandstones are accompanied by purplish shales, which are jointed in a very regular fashion where they are exposed on the northwest side of this bay. Farther round the coast, these shales pass upwards into thin conglomerates, interbedded with sandstones.

Farther northwest along the coast, Easdale Slates make another appearance about 100 yards south of the house at Eilean Orasaig, lying in contact with the sandstones and conglomerates just to the east. They are seen again across another small bay where they are exposed on the promontory, which lies just southwest of the house. These slates are capped by Lower Old Red Sandstone breccias, and the unconformity at the base of these breccias can be traced right round this promontory to the next bay.

Walk up from the shore at this point towards the ruined house at Ardmore, passing another Tertiary dyke which stands up like a wall, 15 feet high, where the Easdale Slates forming its country-rocks have been eroded away. Continue along the path north of Ardmore, crossing Lower Old Red Sandstone conglomerates and sandstones as far as Barnabuck. These rocks pass up into basalt lavas, which form much of the very rough ground west of the path, while these lavas also occur east of the path as an outlier on Torbhain Mor.

Beyond Barnabruck, Easdale Slates are encountered along the track leading back towards Ballliemore and the ferry. Tertiary dykes are seen in various places along this track, together with Lower Old Red Sandstone conglomerates, lying unconformably on top of the Easdale Slates. Just before reaching the jetty, a section along the roadside below the schoolhouse reveals basalt lava with a rubbly zone, marking the top of a lava-flow. Return by ferry to the mainland.

GEOLOGICAL EXCURSION: OBAN TO FINGAL'S CAVE

Daily excursions from Oban visit Fingal's Cave on the island of Staffa during the summer months. Contact the Tourist Information Centre for further details. Leave Oban by boat, sailing north past Kerrera, where the unconformity between Lower Old Red Sandstone conglomerates and black Easdale Slates can be seen at Rubh'a Bhearnaig [NM 841313], together with the raised beach along the northwestern coast of this island. The boat then crosses the Firth of Lorne to the south of Lismore, where outcrops of Dalradian limestone give this island its verdant appearance. The Great Glen Fault runs between Lismore and the mainland of Morvern to the northwest.

The view north towards Morvern is dominated by the rugged terrain underlain by the Strontian Granite at the entrance to the Sound of Mull. This granite is cut off to the northwest by a large fault, running north-south, which reaches the coast at Inninmore Bay. Just beyond this bay, there is a small patch of Carboniferous rocks, overlain by Triassic and Jurassic strata. These rocks are capped by Tertiary lavas, which are exposed farther along the coast to the northwest.

After landing at Craignure on the island of Mull, the coach follows the A849 towards Fionnphort around the Tertiary igneous complex which lies in the centre of Mull. The road first crosses Triassic and Jurassic rocks in the core of the Loch Don Anticline, flanked on either side by Tertiary lavas. However, after turning inland at Loch Spelve, the road crosses the complex and very varied geology formed by the igneous intrusions and pyroclastic rocks of the Tertiary centre on Mull. Like all other Tertiary centres, they represent the roots of an ancient volcano, now revealed by erosion.

Tertiary lavas are again seen on reaching Loch Scridain,, and continue as far west as Bunessan. There are excellent views across this loch, showing the terraced hillsides and escarpments made by the Tertiary lavas, particularly where they are exposed on the slopes of Bearraich. The road then crosses crosses a large fault at Bunessan to reach the Ross of Mull Granite, and its envelope of metamorphic country-rocks, which are exposed at first for about a mile to the west of the fault. The Ross of Mull Granite is a red and rather coarse-grained rock, exposed in smooth but rocky knolls, surrounded by peat, quite unlike the landscape formed by the Tertiary lavas east of Bunessan.

The ferry to Staffa leaves from Fionnphort on the Ross of Mull, overlooking the Sound of Iona. Fingal's Cave is the most celebrated feature of Staffa, penetrating 250 feet into the solid rock with its roof 70 feet above the waves. The columnar jointing of its walls occurs in a basalt lava-flow, which makes the high cliffs riddled with caves at the southern end of the island. The lava-flow rests on top of rather massive agglomerate, which is exposed just above sea-level, west of the landing stage.

Each column can be traced upwards without any break for some 60 feet from the very base of the lava-flow. Above this colonnade lies a jumble of smaller and much less regular columns, forming the upper parts of the lava-flow. Such columnar jointing occurred as the rock cooled down after its eruption as a lava-flow. Elsewhere around Staffa, it takes on a more convoluted form, wherever the lava cooled down in a much less regular fashion. These columnar rocks are very susceptible to erosion by the sea, as the columns fall away very easily from the rest of the rock, forming such features as Fingal's Cave.

OBAN TO TAYVALLICH

Take the A816 south from Oban towards Lochgilphead. The Easdale Slates are exposed at first along the road, before it reaches the outcrop of the Lorne Lavas around Soroba House, just beyond the railway bridge. These volcanic rocks are preserved over a wide area to the east of Oban, giving rise to a low plateau, rarely more than 1,500 feet in height. The landscape typically consists of low knolls and steep-sided crags of rather dark rock, surrounded by boggy ground and the occasional stream. There is a northeast-southwest grain to the country, as the river valleys often follow the trend of igneous dykes, cutting in the same direction across these Devonian lavas.

On reaching Kilninver, a detour can be made along the B844 to *Easdale*, which is the type-locality for the Easdale Slates. These rocks are exposed along the shore at the village of Easdale, overlain unconformably by the Lorne Lavas where they form the cliffs at the back of the raised beach. The slates were once quarried at the village of Easdale, and just offshore on the Island of Easdale. They are cut by Tertiary dykes in several places.

Continuing along the A816 past Kilninver, the main road turns inland at Glen Euchar, following a well-developed terrace of sand and gravel, which can be seen on both sides of the valley. The road then follows the steep-sided valley of Glen Gallain to reach the higher ground overlooking the Pass of Melfort. As it descends towards Kilmelfort, the road crosses the unconformity between the Lorne Lavas and the underlying Dalradian rocks just past a small layby on the new road at [NM 846150]. The rocks lying below this unconformity are the Craignish Phyllites, which can be seen 100 yards farther south along the road, cut by a Tertiary dyke.

The high ground south of Loch Melfort marks the outcrop of the Kilmelfort Granite, which reaches the coast around An Cnap. This granite intrudes the Craignish Phyllites. These slaty rocks give much lower ground, except where they are cut by intrusive bodies of epidiorite. This basic rock is a dolerite which was deformed and metamorphosed under relatively low temperatures at the same time as the Craignish Phyllites took on their present character. Epidiorites make the smooth ridges of higher ground, running northeast-southwest to form the long promontories and low islands, that are such a characteristic feature of this coast. The crags northeast of the road at *Dun an Garbh-Sroine* [NM 803089] provide a typical exposure of epidiorite, making a rather massive, greenish rock, never very dark in colour.

Passing the head of Loch Craignish, the view southwest shows steep slopes facing northwest, and more gentle slopes facing southeast, corresponding to the southeasterly dip of all these Dalradian rocks. They lie to the northwest of the Loch Awe Syncline, which passes down Loch Awe towards Tayvallich and

Knapdale. The road then ascends the steep pass of Bealach Mor, where the Craignish Phyllites give way to the Crinan Grits, lying in the centre of the Loch Awe Syncline. Continue along the road to Kilmartin, where there are several terraces of gravel at different levels along the banks of the Kilmartin Burn.

Turn right a mile beyond Kilmartin along the B8025 road across the very low-lying ground of Moine Mor to reach the Crinan Canal at Bellanoch (weight limit on bridge is only 2 tons). Turn right along the B841, and then fork left along the B8025 towards Tayvallich. Although all this ground is covered in forest, the underlying grain of the country is still northeast-southwest, determining the course of the road as it wriggles its way to Tayvallch. On reaching this village, continue southwest towards Keills, but park as convenient after 1 mile where the track to Upper Fernoch leaves the road at [NR 732860].

GEOLOGICAL EXCURSION: TAYVALLICH PENINSULA

This excursion involves a rather rough walk across the Tayvallich peninsula to the coast of the Sound of Jura. There, the Tayvallich Limestone and Pillow Lavas can be examined southwest of Port Bealach nan Gall. Walk up the track past Upper Fernoch, and then head west across country. Stop near the spot height of 66 metres shown at [NR 723861] on the Ordnance Survey Map (Sheet 55). The ground to the northeast is underlain by the Crinan Grits, folded together with much epidiorite.

These rocks form a rim of slightly higher ground, looking out over a shallow valley to the southwest. This marks the outcrop of the Tayvallich Limestone where it crosses the Tayvallich peninsula, folded around the hinge of the Tayvallich Syncline. The higher ground with its rocky crags of epidiorite, which lies beyond the low-lying ground to the southwest, marks the outcrop of the Tayvallich Pillow Lavas in the core of the syncline.

Walk southwest from this point, following the outcrop of the Tayvallich Limestone across the lower ground towards Port Bealach nan Gall. The Crinan Grits are exposed to the northwest, forming the slightly higher ground opposite Eilean Fraoich. They are best examined on the coast just north of Port Bealach nan Gall. Then walk southwest along the coast after crossing the marshy ground at the head of this bay, crossing epidiorite where it outcrops as a low ridge of rather schistose rock. Continue southwest along the coast to reach a wall at the mouth of a small burn [NR 712852]. The Tayvallich Limestone is exposed beyond this point, lying southeast of the ridge of epidiorite, which continues to make the coast-line.

The Tayvallich Limestone is a dark blue-grey, brown-weathering limestone, often sandy or even quite gritty, dipping towards the southeast. The gritty material is mostly formed by quartz grains, which are often milky-white or bluish in colour with an opalescence appearance. There are similar grains of blue quartz in the Crinan Grits. They are thought to be derived from the Lewisian Gneiss. These limestones display graded bedding where the detrital grains of quartz become finer-grained towards the top of a particular bed, away from its base. It shows that these beds are not overturned, so that the rocks become younger towards the southeast.

Farther along the coast, dolomite makes an appearance in the Tayvallich Limestone, weathering to a very distinctive orangey-brown colour. These dolomites often occur as largish fragments, forming what is a breccia. After about 500 yards, the Tayvallich Limestone runs out to sea, and the Tayvallich Pillow Lavas reach the coast near Port nan Clach Cruinn.

These pillow lavas lie to the southeast of the limestone, and dip steeply in the same direction. Although they consist of epidiorite, appearing much like an intrusive rock, the presence of pillows clearly indicates that these rocks were erupted as lava-flows under water. Admittedly, the pillows are not perfectly displayed, but they can be recognised by the way the outcrops weather into bulbous masses.

There are several lava-flows present along the section to the southwest. The first makes a wall of rock across the foreshore, standing 10 feet in height above the limestone and slates to its northwest. This lava-flow is typical. It has quite an irregular base where tongues of the underlying sediment have penetrated upwards into the lava-flow, separated from one another by bulbous masses of lava. Traced along its strike towards the southwest, the base of this lava-flow appears offset every 15 yards or so, perhaps suggesting that it ploughed into soft sediment as it flowed over the sea-floor.

Other lava-flows are divided by sediment into segments, or else suddenly end along strike. Commonly, the pillows are best developed towards the top of each lava-flow, so that they are most easily seen on its upper surface. The individual flows are usually about 12 feet thick, separated from one another by slates and orangey-brown dolomite.

Return to the road by walking back towards Port Bealach nan Gall, and then following a path which climbs uphill towards the east. Leave the path where it crosses a fence and follow this fence towards Barbreack. On reaching the high ground, turn north and walk a short distance to the viewpoint, looking out over the lower ground to the northeast. This view is the exact opposite to that described on the outward walk.

The crags just below the viewpoint are formed by the Tayvallich Pillow Lavas, lying in the core of the Tayvallich Syncline. The grassy outcrop of the Tayvallich Limestone can be followed around the foot of these crags, folded back on itself by the syncline. The Crinan Grits are exposed in the somewhat higher ground beyond the Tayvallich Limestone, and their outcrop can likewise be traced around this synclinal fold. There is a vivid impression of standing on the bow of a ship, facing away from the southwesterly plunge of the Tayvallich Syncline. Return southwest along the ridge to Barbreack, and then walk east across the lower ground to the road.

Return along the B8025 to Barnaluasgan, and then turn right along the minor road that runs southeast of Loch Sween to Kilmory Bay. After passing

Castle Sween, park in the old village of Kilmory, opposite the ruined chapel. Walk back up the road to the sharp bend at [NR 702753], just west of the village. (Alternatively, walk down the path to the beach at Kilmory Bay to avoid the need for some scrambling over rough terrain at the start of this excursion). On reaching the bend in the road, walk west of northwest along a wall to reach the coast at [NR 697756].

GEOLOGICAL EXCURSION: KILMORY BAY

The Crinan Grits around Kilmory Bay are folded into two synclines, separated from one another by an anticline. These major folds, together with an anticline along Loch Sween and the Tayvallich Syncline even farther to the northwest, make up the Loch Awe Syncline in Knapdale. Pebbly quartzites are exposed on the coast near the end of the wall. The bedding is vertical, striking north of northeast, and graded bedding shows that these rocks become younger towards the east of southeast. The rocks are cut by a slaty cleavage along which the detrital grains in the rock are flattened (MFG 214).

These beds can be followed for 500 yards along the strike towards the south, beyond which they come to dip steeply northeast in the hinge of a major fold. Continue along the coast over rocky outcrops, which require some scrambling, to reach the jetty at [NR 696747], opposite Eilean a'Chapuill. The pebbly quartzites pass downwards into fine-grained quartzites, impure limestones and slates, which are exposed around the low headland to the west of Kilmory Bay. These well-bedded rocks display impressive examples of quite tight folds on a small scale, all plunging steeply to the northeast. This locality can also be reached from the end of the path leading down to Kilmory Bay.

Walk southeast across Kilmory Bay to reach the next exposures. Lying beyond some gritty quartzites are more fine-grained quartzites, impure limestones and slates, all folded together in a very spectacular fashion around [NR 700744]. It is possible at low tide to walk along one particular quartzite, following its course as it is folded back and forth across a closely-packed series of really quite tight folds. These folds plunge at a gentle angle towards the northeast, so that their crests are exposed as rounded surfaces of quartzite, rising gently to the southwest, but falling steeply away on either side, northwest and southeast.

Cross a small stream to reach more pebbly quartzites, which are exposed around [NR 699741]. These rocks dip steeply towards the northwest, while graded bedding in the more conglomeratic horizons shows that they become younger in the same direction. Around a slight headland, they come into contact with the underlying Ardrishaig Phyllites, which are well-exposed beyond a marshy inlet at Port Ban [NR 700740]. Although this formation consists predominantly of slaty rocks, these are folded together with some fine-grained quartzites in a very intricate fashion along the southeast side of Port Ban.

Walk up from the shore at Port Ban to reach the road at Fearnoch, unless planning to continue along the raised beach to the Point of Knap. Although the geology is very complex, there are many features of interest to be seen, while the Point of Knap itself makes a splendid viewpoint over the Sound of Jura.

Index of Geological Terms
and Localities

Page numbers given in bold type refer to entries giving definitions of
geological terms

Agglomerate, **17**

Amphibole.

   See Rock-Forming Minerals

Amphibolite, **20**

Andesite, **19**

Angular Unconformity, 5-6, 15-16,
   23-5, 36, 42, 44, 51, 53-8, 59-60,
   64-5, 67, 94-6

Anticline.

   See Folds and Folding

Argyllshire

   Ardnamurchan, 83

   Easdale, 96

   Falls of Lora, 90

   Fingal's Cave, Staffa, 96

   Ganavan Bay, near Oban, 91-2

   Glencoe, 87-8

   Glenorchy, 89

   Kentallen, 86

   Kerrera, 92-5

   Kilmory Bay, Knap, 99-100

   Oban, 91

   Rubh' a'Bhaid Bheithe, near
      Ballachulish, 86

   Tayvallich, 97-9

Arkose, **12**

Ash (and Volcanic Tuff), **17**

Augite.

   See Rock-Forming Minerals

Basalt, **8**, **19**

Basement-Complex, **20**

Batholiths.

   See Igneous intrusions

Bedding of Sedimentary Rocks,
   7-8, 11-12

Biotite.

   See Rock-forming Minerals

Boulder Clay, **31**

Breccia, **12**

Buried Landscape, **16**, **23**

Caithness

   Clardon Head, 41

   Dunbeath Bypass, 37

   Duncansby Head, 38-9

   Dunnet Head, 40

   Laidhay Croft Museum, 37

   Ness of Duncansby, 38

   Ousdale, 36

   Point of Ness, Dunnet Bay, 40

   Red Point, 41-2

   Scarlet Haven, 37

   South Head of Wick, 37

Calcite, **13**

Caledonian Earth-Movements,
   21, **26**

Caledonian Granites, 28, 36, 41-2,
   83, 84-7, 89, 95-7

Cambrian Quartzite, 25, 45-8, 49-58, 63-7, 70-1.
See also Cambro-Ordovician Rocks

Cambro-Ordovician Rocks, 21, 25, 40, 44-8, 49-58, 63-7, 69-73

Chilled Margins.
See Igneous Rocks (Intrusions)

Columnar Jointing, 8, 17, 76, 77-8, 92, 95-6

Conglomerate, 12

Contact Metamorphism.
See Metamorphic Rocks

Corries, 31

Country-Rocks, 9, 17, 19,
see also Igneous Intrusions, and Metamorphic Rocks

Cross-bedding, 13

Dalradian Schists, 21, 27-28, 84-90, 91-100

Devonian Rocks,
See Old Red Sandstone

Devonian Lavas, 28, 84-5, 87-90, 91, 93, 96.
See also Lorne Lavas (Devonian)

Diorite, 18

Dip (and Strike), 14

Dolerite, 19

Dolomite, 14

Double Unconformity, 25, 54-5

Drumlin, 32

Durness Limestone, 25, 45, 47, 49, 53, 54, 56-8, 63, 67, 70-3.
See also Cambro-Ordovian Rocks

Dyke.
See Igneous Intrusions

Earth-Movements, 6, 14-16

Easter Ross
Black Rock Gorge, 33
Craigton, 33
Hugh Miller's Birthplace, 34
Tarbat Ness, 34

Faults and Faulting, 6, 15
Fault-Plane, 15
Normal Fault, 15
Reverse Fault, 15
Strike-Slip Fault, 15

Feldspar.
See Rock-Forming Minerals

Fluvio-Glacial Sands and Gravels, 32

Folds and Folding, 6, 14
Anticline, 14
Fold-Hinges and Fold-Limbs, 14
Syncline, 14

Fossils, 1-2, 9-10, 25, 29, 34, 36, 56, 69, 76

Fucoid Beds, 25, 46, 52-8, 63, 65, 70.
See also Cambro-Ordovician Rocks

Gabbro, 18

Garnetiferous Mica-Schist, 20

Geological Periods.
See Stratigraphic Column

Geological Time, 9-10

Glacial Erosion and Deposition, 30-2

Glencoe Cauldron-Subsidence, 28, 87-9

Glencoul Thrust, 52-4

Gneiss, 20

Graded-bedding, 13

Granite, 8-9, 18

Granodiorite, 18

Great Glen Fault, 15, 21, 26, 33, 79-81, 83-4, 86, 95

Greywacké, 12

Hornblende-Gneiss, 23

Hornfels, 19

Hutton, James (1726-97), Founder of Modern Geology, 2-9, 11, 15

Igneous Rocks, Nature and Origin, 8-9, 11, 16-19
  See also Pyroclastic Rocks
  Intrusions and Intrusive Rocks, 17
  Lava-Flows and Extrusive Rocks, 8, 16, 27
  Rock-Names and Classification, 18
  Textures and Grain-size, 17

Initial Dip, 8, 16

Inverness-shire
  Ben Nevis, 84-5
  Creag Ghobar, near Loch Eilt, 81
  Glen Roy, 80-1

Loch Quoich, 80

Lochailort, 82

Joints and Jointing, 15

Jurassic Rocks, 21, 29-30, 33, 35-6, 69-72, 74-8, 95

Landslips, 76, 78

Lava-Flows.
  See Igneous Rocks

Laxfordian Gneisses, 23, 47, 50-1

Lewisian Gneiss, 21, 22-25, 40, 44-48, 49-58, 59-62, 63-68, 69-71, 79.
  See also Laxfordian and Scourian Gneisses

Limestone, 13

Lorne Lavas (Devonian), 21, 28, 84, 90-1, 93, 96

Magma.
  See Igneous Rocks, Nature and Origin

Marble, 19

Metamorphic Rocks, Nature and Origin, 9, 11, 19-20
  Aureoles of Thermal Metamorphism, 17
  Contact or Thermal Metamorphism, 19
  Rock-Names and Classification, 19-20
  Regional Metamorphism, 19
  Thermal or Contact Metamorphism, 19

Mica.
See Rock-Forming Minerals
Migmatite, 28
Moine Schists, 21, 26-7, 33, 37, 40,
43-4, 46-7, 49, 52, 57, 63-4, 68,
79-83
Moine Thrust, 21, 26, 40, 44-9, 52,
56-8, 63-8, 69-70, 79
Moraines, 32
Mountain-Building.
See Earth-Movements
Mudstone, 13
Muscovite.
See Rock-Forming Minerals
Mylonite, 20

Neptunist Theory of Werner, 7
New Red Sandstone.
See Permo-Triassic rocks
Normal Fault.
See Faults and Faulting

Old Red Sandstone (Devonian),
21, 28-9, 33-7, 40-3, 90-5
Olivine.
See Rock-Forming Minerals

Parallel Roads of Glen Roy, 79-80
Pegmatite, 18
Permo-Triassic Rocks, 21, 29, 33,
35, 44, 69, 71, 95
Phenocrysts.
See Igneous Rocks (Textures)
Phyllite, 20

Pillow-Lavas, 27, 98-9
Pipe Rock, 25, 44-7, 49-58, 63,
65-7, 70
See also Cambro-Ordovician
Rocks
Porphyritic Texture.
See Igneous Rocks (Textures)
Portgower Boulder-Beds, 29, 35-6
Pyroclastic rocks, Nature and
Origin, 17
Pyroxene.
See Rock-Forming Minerals
Pyroxene-Granulites, 23

Quartz.
See Rock-Forming Minerals
Quartz-Diorite, 18
Quartzite (Sedimentary), 12
Quartzite (Metamorphic), 19

Raised Beaches, 32
Regional Metamorphism.
See Metamorphic rocks, Nature
and Origin
Reverse Fault.
See Faults and Faulting
Rhyolite, 19
Roches Moutonnée, 31
Rock-Forming Minerals, 17-18

Sandstone, 12
Schist, 20
Schistosity, 20
Scourian Gneisses, 23, 51

Scourie Dykes, 22-3, 50-1

Sedimentary Rocks, Nature and
Origin, 3-4, 11-14
See also Bedding of Sedimentary
Rocks
Consolidation (Compaction and
Cementation), 6
Principle of Superposition, 9, 12
Rock-Names and Classification,
12-14
Stratigraphic Dating, 10

Serpulite Grit, 25, 46, 54, 56-7,
65, 70.
See also Cambro-Ordovician
Rocks

Shale, 13

Sill.
See Igneous Intrusions

Skye
Ardnish, near Broadford, 70
Berreraig Bay, 76
Camas Ban, near Portree, 76
Camas Malag, near Torrin, 71
Kilt Rock, 77
Loch Cill Chriosd, 73-4
Loch Coruisk, 72
Loch Slapin, 72
Old Man of Storr, 76-7
Ord, 70-1
Sligachan, 75
The Quiraing, 77-8

Slate and Slaty Cleavage, 20

Stoer Group, 24, 59, 61.
See also Torridonian Sandstone

Stratification.
See Bedding of Sedimentary
Rocks

Strike (and Dip), 14

Strike-Slip Fault.
See Faults and Faulting

Sutherland
Beinn Ceannabeinne, 46
Ben Arnaboll, 45
Ben Heilam, 45
Ben Hutig, 44
Cape Wrath, 48
Ceannabeinne, near Durness, 46
Clachtoll, 59-60
Cnoc Mor, near Kirtomy, 43
Coldbackie Bay, 43
Gualin House, 49
Kinloch, near Tongue, 44
Laxford Brae, 50
Loch Assynt, 53-4
Loch Eriboll, 45-6
Loch Fleet, 35
Loch Glencoul, 52-3
Loch Stack, 51
Loch Vasgo, near Talmine, 44
Orcadian Stone Company,
Golspie, 35
Polla, near Loch Eriboll, 46
Portgower, 35-6
Portskerra, 42
Rispond, 46
Sango Bay, near Durness, 47
Scourie, 51
Skiag Bridge, 54-6
Smoo Cave, 47
Strath of Kildonan, 36
Tarbet, near Scourie, 51
The Mound, 35
Timespan Heritage Centre,
Helmsdale, 36
Torrisdale Bay, 43

Traligill River, near
  Inchnadamph, 56
Upper Badcall, near Scourie, 51
Syncline.
  See Folds and Folding

Tertiary Igneous Rocks, 21, 29-30
  Dykes, 70, 72-3, 76, 82-3, 93-6
  Gabbros, 69, 72-5, 82-3
  Granites, 69, 71-5
  Lavas, 69, 72-3, 75-8, 82, 95-6
  Sills, 76-8
Thermal Metamorphism.
  See Metamorphic Rocks
Torridonian Sandstone
  (Stoer and Torridon Groups),
  21, 23-4, 40, 48-58, 59-62,
  63-68, 69-72, 75, 79
Triassic Rocks
  See Permo-Triassic Rocks
Tuff (and Volcanic Ash), 17

Uniformitarianism, 2, 4-5, 11

Werner, Abram Gottlieb
  (1750-1817), 6-9
Wester Ross
  Aird of Coigach, 60
  Corrie Point, near Ullapool, 63
  Corrieshalloch Gorge, 63
  Dundonell House, 64
  Enard Bay, 61-2
  Glenshiel, 79
  Knockan Cliff, near Elphin, 57-8
  Loch Cluanie, 80
  Loch Maree and Slioch, 65
  Loch Torridon, 67
  Meall a'Ghiubhais, near
    Kinlochewe, 65
  Plockton, 68

Some other books published by **LUATH** PRESS

LUATH GUIDES TO SCOTLAND

'Gentlemen, We have just returned from a six week stay in Scotland. I am convinced that Tom Atkinson is the best guidebook author I have ever read, about any place, any time.'
Edward Taylor, LOS ANGELES

These guides are not your traditional where-to-stay and what-to-eat books. They are companions in the rucksack or car seat, providing the discerning traveller with a blend of fiery opinion and moving description. Here you will find *that curious pastiche of myths and legend and history that the Scots use to describe their heritage... what battle happened in which glen between which clans; where the Picts sacrificed bulls as recently as the 17th century... A lively counterpoint to the more standard, detached guidebook... Intriguing.'*
THE WASHINGTON POST

These are perfect guides for the discerning visitor or resident to keep close by for reading again and again, written by authors who invite you to share their intimate knowledge and love of the areas covered.

South West Scotland

Tom Atkinson
ISBN 0 946487 04 9 PBK £4.95

This descriptive guide to the magical country of Robert Burns covers Kyle, Carrick, Galloway, Dumfries-shire, Kirkcudbrightshire and Wigtownshire. Hills, unknown moors and unspoiled

beaches grace a land steeped in history and legend and portrayed with affection and deep delight.

An essential book for the visitor who yearns to feel at home in this land of peace and grandeur.

The Lonely Lands

Tom Atkinson
ISBN 0 946487 10 3 PBK £4.95

A guide to Inveraray, Glencoe, Loch Awe, Loch Lomond, Cowal, the Kyles of Bute and all of central Argyll written with insight, sympathy and loving detail. Once Atkinson has taken you there, these lands can never feel lonely. 'I have sought to make the complex simple, the beautiful accessible and the strange familiar,' he writes, and indeed he brings to the land a knowledge and affection only accessible to someone with intimate knowledge of the area.

A must for travellers and natives who want to delve beneath the surface.

'Highly personal and somewhat quirky... steeped in the lore of Scotland.'
THE WASHINGTON POST

The Empty Lands

Tom Atkinson
ISBN 0 946487 13 8 PBK £4.95

The Highlands of Scotland from Ullapool to Bettyhill and Bonar Bridge to John O'Groats are landscapes of myth and legend, 'empty of people, but of nothing else that brings delight to any tired soul,' writes Atkinson. This highly personal guide describes Highland his-

LUATH PRESS LIMITED

tory and landscape with love, compassion and above all sheer magic.
Essential reading for anyone who has dreamed of the Highlands.

Roads to the Isles

Tom Atkinson

ISBN 0 946487 01 4 PBK £4.95

Ardnamurchan, Morvern, Morar, Moidart and the west coast to Ullapool are included in this guide to the Far West and Far North of Scotland. An unspoiled land of mountains, lochs and silver sands is brought to the walker's toe-tips (and to the reader's fingertips) in this stark, serene and evocative account of town, country and legend.

For any visitor to this Highland wonderland, Queen Victoria's favourite place on earth.

Highways and Byways in Mull and Iona

Peter Macnab

ISBN 0 946487 16 2 PBK £4.25

'The Isle of Mull is of Isles the fairest, Of ocean's gems 'tis the first and rarest.' So a local poet described it a hundred years ago, and this recently revised guide to Mull and sacred Iona, the most accessible islands of the Inner Hebrides, takes the reader on a delightful tour of these rare ocean gems, travelling with a native whose unparalleled knowledge and deep feeling for the area unlock the byways of the islands in all their natural beauty.

NATURAL SCOTLAND

Rum: Nature's Island

Magnus Magnusson

ISBN 0 946487 32 4 £7.95 PBK

Rum: Nature's Island is the fascinating story of a Hebridean island from the earliest times through to the Clearances and its period as the sporting playground of a Lancashire industrial magnate, and on to its rebirth as a National Nature Reserve, a model for the active ecological management of Scotland's wild places.

Thoroughly researched and written in a lively accessible style, the book includes comprehensive coverage of the island's geology, animals and plants, and people, with a special chapter on the Edwardian extravaganza of Kinloch Castle. There is practical information for visitors to what was once known as 'the Forbidden Isle'; the book provides details of bothy and other accommodation, walks and nature trails. It closes with a positive vision for the island's future: biologically diverse, economically dynamic and ecologically sustainable.

Rum: Nature's Island is published in co-operation with Scottish Natural Heritage (of which Magnus Magnusson is Chairman) to mark the 40th anniversary of the acquisition of Rum by its predecessor, The Nature Conservancy.

Wild Scotland: The essential guide to finding the best of natural Scotland

James McCarthy

Photography by Laurie Campbell

ISBN 0 946487 37 5 PBK £7.50

With a foreword by Magnus Magnusson and striking colour photographs by

Laurie Campbell, this is the essential up-to-date guide to viewing wildlife in Scotland for the visitor and resident alike. It provides a fascinating overview of the country's plants, animals, bird and marine life against the background of their typical natural settings, as an introduction to the vivid descriptions of the most accessible localities, linked to clear regional maps. A unique feature is the focus on 'green tourism' and sustainable visitor use of the countryside, contributed by Duncan Bryden, manager of the Scottish Tourist Board's Tourism and the Environment Task Force. Important practical information on access and the best times of year for viewing sites makes this an indispensable and user-friendly travelling companion to anyone interested in exploring Scotland's remarkable natural heritage.

James McCarthy is former Deputy Director for Scotland of the Nature Conservancy Council, and now a Board Member of Scottish Natural Heritage and Chairman of the Environmental Youth Work National Development Project Scotland.

An Inhabited Solitude: Scotland – Land and People

James McCarthy

Photography by Laurie Campbell

ISBN 0 946487 30 8 PBK £6.99

'Scotland is the country above all others that I have seen, in which a man of imagination may carve out his own pleasures; there are so many inhabited solitudes.'

DOROTHY WORDSWORTH, in her journal of August 1803

An informed and thought-provoking profile of Scotland's unique landscapes and the impact of humans on what we see now and in the future, James McCarthy leads us through the many aspects of the land and the people who inhabit it: natural Scotland; the rocks beneath; land ownership; the use of resources; people and place; conserving Scotland's heritage and much more.

Written in a highly readable style, this concise volume offers an understanding of the land as a whole. Emphasising the uniqueness of the Scottish environment, the author explores the links between this and other aspects of our culture as a key element in rediscovering a modern sense of the Scottish identity and perception of nationhood.

'This book provides an engaging introduction to the mysteries of Scotland's people and landscapes. Difficult concepts are described in simple terms, providing the interested Scot or tourist with an invaluable overview of the country... It fills an important niche which, to my knowledge, is filled by no other publications.'

BETSY KING, Chief Executive, Scottish Environmental Education Council.

WALK WITH LUATH

Mountain Days & Bothy Nights

Dave Brown and Ian Mitchell

ISBN 0 946487 15 4 PBK £7.50

Acknowledged as a classic of mountain writing still in demand ten years after its first publication, this book takes you into the bothies, howffs and dosses on the Scottish hills. Fishgut Mac, Desperate Dan and Stumpy the Big Yin stalk hill and public house, evading gamekeepers and Royalty with a camaraderie which was the trademark of Scots hill-walking in the early days.

'*The fun element comes through… how innocent the social polemic seems in our nastier world of today… the book for the rucksack this year.*'

Hamish Brown, SCOTTISH MOUNTAINEERING CLUB JOURNAL

'*The doings, sayings, incongruities and idiosyncrasies of the denizens of the bothy underworld… described in an easy philosophical style… an authentic word picture of this part of the climbing scene in latter-day Scotland, which, like any good picture, will increase in charm over the years.*'

Iain Smart, SCOTTISH MOUNTAINEERING CLUB JOURNAL

'*The ideal book for nostalgic hillwalkers of the 60s, even just the armchair and public house variety… humorous, entertaining, informative, written by two men with obvious expertise, knowledge and love of their subject.*'

SCOTS INDEPENDENT

'*Fifty years have made no difference. Your crowd is the one I used to know… [This] must be the only complete dossers' guide ever put together.*'

Alistair Borthwick, author of the immortal *Always a Little Further*.

The Joy of Hillwalking

Ralph Storer

ISBN 0 946487 28 6 PBK £7.50

Apart, perhaps, from the joy of sex, the joy of hillwalking brings more pleasure to more people than any other form of human activity.

'*Alps, America, Scandinavia, you name it – Storer's been there, so why the hell shouldn't he bring all these various and varied places into his observations… [He] even admits to losing his virginity after a*

day on the Aggy Ridge… Well worth its place alongside Storer's earlier works.'
TAC

Scotland's Mountains before the Mountaineers

Ian Mitchell

ISBN 0 946487 39 1 PBK £9.99

How many Munros did Bonnie Prince Charlie bag?

Which clergyman climbed all the Cairngorm 4,000-ers nearly two centuries ago?

Which bandit and sheep rustler hid in the mountains while his wife saw off the sheriff officers with a shotgun?

According to Gaelic tradition, how did an outlier of the rugged Corbett Beinn Aridh Charr come to be called Spidean Moirich, 'Martha's Peak'?

Who was the murderous clansman who gave his name to Beinn Fhionnlaidh?

In this ground-breaking book, Ian Mitchell tells the story of explorations and ascents in the Scottish Highlands in the days before mountaineering became a popular sport – when bandits, Jacobites, poachers and illicit distillers traditionally used the mountains as sanctuary. The book also gives a detailed account of the map makers, road builders, geologists, astronomers and naturalists, many of whom ascended hitherto untrodden summits while working in the Scottish Highlands.

Scotland's Mountains before Mountaineers is divided into four Highland regions, with a map of each region showing key summits. While not designed primarily as a guide, it will be a useful handbook for walkers and climbers. Based on a wealth of new research, this book offers a fresh perspective that will fascinate climbers and mountaineers and everyone interested in the history of mountaineering, cartography, the evolution of landscape and the social history of the Scottish Highlands.

Cairngorm region. Here is the answer: rambles through scenic woods with a welcoming pub at the end, birdwatching hints, glacier holes, or for the fit and ambitious, scrambles up hills to admire vistas of glorious scenery. Wildlife in the Cairngorms is unequalled elsewhere in Britain, and here it is brought to the binoculars of any walker who treads quietly and with respect.

LUATH WALKING GUIDES

The highly respected and continually updated guides to the Cairngorms.

'Particularly good on local wildlife and how to see it'
THE COUNTRYMAN

Walks in the Cairngorms

Ernest Cross
ISBN 0 946487 09 X PBK £3.95

This selection of walks celebrates the rare birds, animals, plants and geological wonders of a region often believed difficult to penetrate on foot. Nothing is difficult with this guide in your pocket, as Cross gives a choice for every walker, and includes valuable tips on mountain safety and weather advice.

Ideal for walkers of all ages and skiers waiting for snowier skies.

Short Walks in the Cairngorms

Ernest Cross
ISBN 0 946487 23 5 PBK £3.95

Cross wrote this volume after overhearing a walker remark that there were no short walks for lazy ramblers in the

SPORT

Over the Top with the Tartan Army (Active Service 1992-97)

Andrew McArthur
ISBN 0 946487 45 6 PBK £7.99

Scotland has witnessed the growth of a new and curious military phenomenon - grown men bedecked in tartan yomping across the globe, hell-bent on benevolence and ritualistic bevvying. What noble cause does this famous army serve? Why, football of course!

Taking us on an erratic world tour, McArthur gives a frighteningly funny insider's eye view of active service with the Tartan Army - the madcap antics of Scotland's travelling support in the '90s, written from the inside, covering campaigns and skirmishes from Euro '92 up to the qualifying drama for France '98 in places as diverse as Russia, the Faroes, Belarus, Sweden, Monte Carlo, Estonia, Latvia, USA and Finland.

This book is a must for any football fan who likes a good laugh.

'I commend this book to all football supporters'. Graham Spiers, SCOTLAND ON SUNDAY

'In wishing Andy McArthur all the best

with this publication, I do hope he will be in a position to produce a sequel after our participation in the World Cup in France!'
CRAIG BROWN, Scotland Team Coach

All royalties on sales of the book are going to Scottish charities, principally Children's Hospice Association Scotland, the only Scotland-wide charity of its kind, providing special love and care to children with terminal illnesses at its hospice, Rachel House, in Kinross.

Ski & Snowboard Scotland

Hilary Parke
ISBN 0 946487 35 9 PBK £6.99

How can you cut down the queue time and boost the snow time?

Who can show you how to cannonball the quarterpipe?

Where are the bumps that give most air-time?

Where can you watch international rugby in-between runs on the slopes?

Which mountain restaurant serves magical Mexican meals?

Which resort has the steepest on-piste run in Scotland?

Where can you get a free ski guiding service to show you the best runs?

If you don't know the answers to all these questions - plus a hundred or so more then this book is for you!

Snow sports in Scotland are still a secret treasure. There's no need to go abroad when there's such an exciting variety of terrain right here on your doorstep. You just need to know what to look for. Ski & Snowboard Scotland is aimed at maximising the time you have available so that the hours you spend on the snow are memorable for all the right reasons. This fun and informative book guides

you over the slopes of Scotland, giving you the inside track on all the major ski centres. There are chapters ranging from how to get there to the impact of snowsports on the 'environment.

"Reading the book brought back many happy memories of my early training days at the dry slope in Edinburgh and of many brilliant weekends in the Cairngorms.'

EMMA CARRICK-ANDERSON, from her foreword, written in the US, during a break in training for her first World Cup as a member of the British Alpine Ski Team.

SOCIAL HISTORY

The Crofting Years

Francis Thompson
ISBN 0 946487 06 5 PBK £6.95

Crofting is much more than a way of life. It is a storehouse of cultural, linguistic and moral values which holds together a scattered and struggling rural population. This book fills a blank in the written history of crofting over the last two centuries. Bloody conflicts and gunboat diplomacy, treachery, compassion, music and story: all figure in this mine of information on crofting in the Highlands and Islands of Scotland.

'I would recommend this book to all who are interested in the past, but even more so to those who are interested in the future survival of our way of life and culture'
STORNOWAY GAZETTE

'A cleverly planned book... the story told in simple words which compel attention... [by] a Gaelic speaking Lewisman with specialised knowledge of the crofting community.'
BOOKS IN SCOTLAND

'The book is a mine of information on many aspects of the past, among them the homes, the food, the music and the medicine of our crofting forebears.'
John M Macmillan, erstwhile CROFTERS COMMISSIONER FOR LEWIS AND HARRIS

'This fascinating book is recommended to anyone who has the interests of our language and culture at heart.'
Donnie Maclean, DIRECTOR OF AN COMUNN GAIDHEALACH, WESTERN ISLES

'Unlike many books on the subject, Crofting Years combines a radical political approach to Scottish crofting experience with a ruthless realism which while recognising the full tragedy and difficulty of his subject never descends to sentimentality or nostalgia'
CHAPMAN

MUSIC AND DANCE

Highland Balls and Village Halls

GW Lockhart
ISBN 0 946487 12 X PBK £6.95
Acknowledged as a classic in Scottish dancing circles throughout the world. Anecdotes, Scottish history, dress and dance steps are all included in this

'delightful little book, full of interest... both a personal account and an understanding look at the making of traditions.'
NEW ZEALAND SCOTTISH COUNTRY DANCES MAGAZINE

'A delightful survey of Scottish dancing and custom. Informative, concise and opinionated, it guides the reader across the history and geography of country dance and ends by detailing the 12 dances every Scot should know – the most famous being the Eightsome Reel, "the greatest longest, rowdiest, most diabolically executed of all the Scottish country dances".'
THE HERALD

'A pot-pourri of every facet of Scottish country dancing. It will bring back memories of petronella turns and poussettes and make you eager to take part in a Broun's reel or a dashing white sergeant!'
DUNDEE COURIER AND ADVERTISER

'An excellent an very readable insight into the traditions and customs of Scottish country dancing. The author takes us on a tour from his own early days jigging in the village hall to the characters and traditions that have made our own brand of dance popular throughout the world.'
SUNDAY POST

Fiddles & Folk: A celebration of the re-emergence of Scotland's musical heritage

GW Lockhart
ISBN 0 946487 38 3 PBK £7.99
In Fiddles & Folk, his companion volume to Highland Balls and Village Halls, now an acknowledged classic on Scottish dancing, Wallace Lockhart meets up with many of the people who have created the renaissance of Scotland's music at home and overseas. From Dougie MacLean, Hamish Henderson, the Battlefield Band, the Whistlebinkies, the Scottish Fiddle Orchestra, the McCalmans and many more come the stories that break down the musical barriers between Scotlandís past and present, and between the diverse musical forms which have woven together to create the dynamism of the music today.

'I have tried to avoid a formal approach to Scottish music as it affects those of us with our musical heritage coursing through our veins. The picture I have sought is one of *many brush strokes, looking at how some individuals have come to the fore, examining their music, lives, thoughts, even philosophies...'* WALLACE LOCKHART

' *"I never had a narrow, woolly-jumper, fingers stuck in the ear approach to music. We have a musical heritage here that is the envy of the rest of the world. Most countries just can't compete,"* he *[Ian Green, Greentrax] says. And as young Scots tire of Oasis and Blur, they will realise that there is a wealth of young Scottish music on their doorstep just waiting to be discovered.'* THE SCOTSMAN, March 1998

For anyone whose heart lifts at the sound of fiddle or pipes, this book takes you on a delightful journey, full of humour and respect, in the company of some of the performers who have taken Scotland's music around the world and come back enriched.

FOLKLORE

The Supernatural Highlands

Francis Thompson

ISBN 0 946487 31 6 PBK £8.99

An authoritative exploration of the otherworld of the Highlander, happenings and beings hitherto thought to be outwith the ordinary forces of nature. A simple introduction to the way of life of rural Highland and Island communities, this new edition weaves a path through second sight, the evil eye, witchcraft, ghosts, fairies and other supernatural beings, offering new sight-lines on areas of belief once dismissed as folklore and superstition.

Tall Tales from an Island

Peter Macnab

ISBN 0 946487 07 3 PBK £8.99

Peter Macnab was born and reared on Mull. He heard many of these tales as a lad, and others he has listened to in later years. Although collected on Mull, they could have come from any one of the Hebridean islands. Timeless and universal, these tales are still told round the fireside when the visitors have all gone home.

There are humorous tales, grim tales, witty tales, tales of witchcraft, tales of love, tales of heroism, tales of treachery, historical tales and tales of yesteryear. There are unforgettable characters like Do'l Gorm, the philosophical roadman, and Calum nan Croig, the Gaelic storyteller whose highly developed art of convincing exaggeration mesmerised his listeners. There is a headless horseman, and a whole coven of witches. Heroes, fools, lairds, herdsmen, lovers and liars, dead men and live cats all have a place in this entrancing collection. This is a superb collection indeed, told by a master storyteller with all the rhythms remembered from the firesides of his childhood.

A popular lecturer, broadcaster and writer, Peter Macnab is the author of a number of books and articles about Mull, the island he knows so intimately and loves so much. As he himself puts it in his introduction to this book 'I am of the unswerving opinion that nowhere else in the world will you find a better way of life, nor a finer people with whom to share it.'

'All islands, it seems, have a rich store of characters whose stories represent a kind of

sub-culture without which island life would be that much poorer. Macnab has succeeded in giving the retelling of the stories a special Mull flavour, so much so that one can visualise the storytellers sitting on a bench outside the house with a few cronies, puffing on their pipes and listening with nodding approval.' WEST HIGHLAND FREE PRESS

FICTION

The Bannockburn Years

William Scott

ISBN 0 946487 34 0 PBK £7.95

A present day Edinburgh solicitor stumbles across reference to a document of value to the Nation State of Scotland. He tracks down the document on the Isle of Bute, a document which probes the real 'quaestiones' about nationhood and national identity. The document ends up being published, but is it authentic and does it matter? Almost 700 years on, these 'quaestiones' are still worth asking.

Written with pace and passion, William Scott has devised an intriguing vehicle to open up new ways of looking at the future of Scotland and its people. He presents an alternative interpretation of how the Battle of Bannockburn was fought, and through the Bannatyne manuscript he draws the reader into the minds of those involved.

Winner of the 1997 Constable Trophy, the premier award in Scotland for an unpublished novel, this book offers new insights to both the academic and the general reader which are sure to provoke further discussion and debate.

'A brilliant storyteller. I shall expect to see your name writ large hereafter.' NIGEL TRANTER, October 1997.

'... a compulsive read'. PH SCOTT, THE SCOTSMAN

The Great Melnikov

Hugh MacLachlan

ISBN 0 946487 42 1 PBK £7.95

A well crafted, gripping novel, written in a style reminiscent of John Buchan and set in London and the Scottish Highlands during the First World War, *The Great Melnikov* is a dark tale of double-cross and deception. We first meet Melnikov, one-time star of the German circus, languishing as a down-and-out in Trafalgar Square. He soon finds himself drawn into a tortuous web of intrigue. He is a complex man whose personal struggle with alcoholism is an inner drama which parallels the tense twists and turns as a spy mystery unfolds. Melnikov's options are narrowing. The circle of threat is closing. Will Melnikov outwit the sinister enemy spy network? Can he summon the will and the wit to survive?

Hugh MacLachlan, in his first full length novel, demonstrates an undoubted ability to tell a good story well. His earlier stories have been broadcast on Radio Scotland, and he has the rare distinction of being shortlisted for the Macallan/Scotland on Sunday Short Story Competition two years in succession.

BIOGRAPHY

Tobermory Teuchter: A first-hand account of life on Mull in the early years of the 20th century

Peter Macnab

ISBN 0 946487 41 3 PBK £7.99

Peter Macnab was reared on Mull, as was his father, and his grandfather before him. In this book he provides a revealing account of life on Mull during the first quarter of the 20th century, focusing especially on the years of World War I. This enthralling social history of the island is set against Peter Macnab's early years as son of the governor of the Mull Poorhouse, one of the last in the Hebrides, and is illustrated throughout by photographs from his exceptional collection. Peter Macnab's 'fisherman's yarns' and other personal reminiscences are told delightfully by a born storyteller.

This latest work from the author of a range of books about the island, including the standard study of Mull and Iona, reveals his unparalleled knowledge of and deep feeling for Mull and its people. After his long career with the Clydesdale Bank, first in Tobermory and later on the mainland, Peter, now 94, remains a teuchter at heart, proud of his island heritage.

'Peter Macnab is a man of words who does-nit mince his words - not where his beloved Mull is concerned. 'I will never forget some of the inmates of the poorhouse,' says Peter. 'Some of them were actually victims of the later Clearances. It was history at first hand, and there was no romance about it'. But Peter Macnab sees little cre-ative point in crying over ancient injus-tices. For him the task is to help Mull in this century and beyond.'

SCOTS MAGAZINE, May 1998

Bare Feet and Tackety Boots

Archie Cameron

ISBN 0 946487 17 0 PBK £7.95

The island of Rum before the First World War was the playground of its rich absentee landowner. A survivor of life a century gone tells his story. Factors and schoolmasters, midges and poach-ing, deer, ducks and MacBrayne's steamers: here social history and per-sonal anecdote create a record of a way of life gone not long ago but already almost forgotten. This is the story the gentry couldn't tell.

'This book is an important piece of social history, for it gives an insight into how the other half lived in an era the likes of which will never be seen again'
FORTHRIGHT MAGAZINE

'The authentic breath of the pawky, country-wise estate employee.'
THE OBSERVER

'Well observed and detailed account of island life in the early years of this century'
THE SCOTS MAGAZINE

'A very good read with the capacity to make the reader chuckle. A very talented writer.'
STORNOWAY GAZETTE

On the Trail of Robert Service

GW Lockhart

ISBN 0 946487 24 3 PBK £7.99

Robert Service is famed world-wide for his eye-witness verse-pictures of the Klondike goldrush. As a war poet, his work outsold Owen and Sassoon, and he went on to become the world's first million selling poet. In search of adven-ture and new experiences, he emigrated from Scotland to Canada in 1890 where he was caught up in the aftermath of the raging gold fever. His vivid dramatic verse bring to life the wild, larger than life characters of the gold rush Yukon, their bar-room brawls, their lust for gold, their trigger-happy gambles with life and love. 'The Shooting of Dan McGrew' is perhaps his most famous poem:

A bunch of the boys were whooping it up in the Malamute saloon;
The kid that handles the music box was hitting a ragtime tune;
Back of the bar in a solo game, sat Dangerous Dan McGrew,
And watching his luck was his light o'love, the lady that's known as Lou.

His storytelling powers have brought Robert Service enduring fame, particularly in North America and Scotland where he is something of a cult figure. Starting in Scotland, *On the Trail of Robert Service* follows Service as he wanders through British Columbia, Oregon, California, Mexico, Cuba, Tahiti, Russia, Turkey and the Balkans, finally 'settling' in France.

This revised edition includes an expanded selection of illustrations of scenes from the Klondike as well as several more photographs from the family of Robert Service on his travels around the world. Wallace Lockhart, an expert on Scottish traditional folk music and dance, is the author of *Highland Balls & Village Halls* and *Fiddles & Folk*. His relish for a well-told tale in popular vernacular led him to fall in love with the verse of Robert Service and write his biography.

'*A fitting tribute to a remarkable man - a bank clerk who wanted to become a cowboy. It is hard to imagine a bank clerk writing such lines as:*

A bunch of boys were whooping it up...
The income from his writing actually exceeded his bank salary by a factor of five and he resigned to pursue a full time writing career.' Charles Munn,
THE SCOTTISH BANKER

'*Robert Service claimed he wrote for those who wouldnit be seen dead reading poetry.*

His was an almost unbelievably mobile life... Lockhart hangs on breathlessly, enthusiastically unearthing clues to the poet's life.' Ruth Thomas,
SCOTTISH BOOK COLLECTOR

'*This enthralling biography will delight Service lovers in both the Old World and the New.*' Marilyn Wright,
SCOTS INDEPENDENT

Come Dungeons Dark

John Taylor Caldwell
ISBN 0 946487 19 7 PBK £6.95

Glasgow anarchist Guy Aldred died with 10p in his pocket in 1963 claiming there was better company in Barlinnie Prison than in the Corridors of Power. 'The Red Scourge' is remembered here by one who worked with him and spent 27 years as part of his turbulent household, sparring with Lenin, Sylvia Pankhurst and others as he struggled for freedom for his beloved fellow-man.

'*The welcome and long-awaited biography of... one of this country's most prolific radical propagandists... Crank or visionary?... whatever the verdict, the Glasgow anarchist has finally been given a fitting memorial.*'
THE SCOTSMAN

POETRY

Blind Harry's Wallace

William Hamilton of Gilbertfield
ISBN 0 946487 43 X HBK £15.00
ISBN 0 946487 33 2 PBK £7.50

The original story of the real braveheart, Sir William Wallace. Racy, blood on every page, violently anglophobic, grossly embellished, vulgar and disgusting, clumsy and stilted, a literary failure, a great epic.

Whatever the verdict on BLIND HARRY, this is the book which has done more than any other to frame the notion of Scotland's national identity. Despite its numerous 'historical inaccuracies', it remains the principal source for what we now know about the life of Wallace.

The novel and film *Braveheart* were based on the 1722 Hamilton edition of this epic poem. Burns, Wordsworth, Byron and others were greatly influenced by this version 'wherein the old obsolete words are rendered more intelligible', which is said to be the book, next to the Bible, most commonly found in Scottish households in the eighteenth century. Burns even admits to having 'borrowed... a couplet worthy of Homer' directly from Hamilton's version of BLIND HARRY to include in 'Scots wha hae'.

Elspeth King, in her introduction to this, the first accessible edition of BLIND HARRY in verse form since 1859, draws parallels between the situation in Scotland at the time of Wallace and that in Bosnia and Chechnya in the 1990s. Seven hundred years to the day after the Battle of Stirling Bridge, the 'Settled Will of the Scottish People' was expressed in the devolution referendum of 11 September 1997. She describes this as a landmark opportunity for mature reflection on how the nation has been shaped, and sees BLIND HARRY'S WALLACE as an essential and compelling text for this purpose.

'Builder of the literary foundations of a national hero-cult in a free and powerful country'.
ALEXANDER STODDART, sculptor
'A true bard of the people'
TOM SCOTT, THE PENGUIN BOOK OF SCOTTISH VERSE, on Blind Harry.

'A more inventive writer than Shakespeare'
RANDALL WALLACE

'The story of Wallace poured a Scottish prejudice in my veins which will boil along until the floodgates of life shut in eternal rest'
ROBERT BURNS

'Hamilton's couplets are not the best poetry you will ever read, but they rattle along at a fair pace. In re-issuing this work, the publishers have re-opened the spring from which most of our conceptions of the Wallace legend come'.
SCOTLAND ON SUNDAY

'The return of Blind Harry's Wallace, a man who makes Mel look like a wimp'.
THE SCOTSMAN

Poems to be Read Aloud

Collected and with an introduction by Tom Atkinson
ISBN 0 946487 00 6 PBK £5.00

This personal collection of doggerel and verse ranging from the tear-jerking *Green Eye of the Yellow God* to the rarely printed, bawdy *Eskimo Nell* has a lively cult following. Much borrowed and rarely returned, this is a book for reading aloud in very good company, preferably after a dram or twa. You are guaranteed a warm welcome if you arrive at a gathering with this little volume in your pocket.

This little book is an attempt to stem the great rushing tide of canned entertainment. A hopeless attempt of course. There is poetry of very high order here, but there is also some fearful doggerel. But that is the way of things. No literary axe is being ground..

Of course some of the items in this book are poetic drivel, if read as poems. But that is not the point. They all spring to life when they are read aloud. It is the combination of the poem with your voice, with all the art and craft you can muster, that produces the finished product and effect you seek.

You don't have to learn the poems. Why clutter up your mind with rubbish? Of course, it is a poorly furnished mind that doesn't carry a fair stock of poetry, but surely the poems to be remembered and savoured in secret, when in love, or ill, or sad, are not the ones you want to share with an audience.

So go ahead, clear your throat and transfix all talkers with a stern eye, then let rip!
TOM ATKINSON

Luath Press Limited

committed to publishing well written books worth reading

LUATH PRESS takes its name from Robert Burns, whose little collie Luath (*Gael.*, swift or nimble) tripped up Jean Armour at a wedding and gave him the chance to speak to the woman who was to be his wife and the abiding love of his life. Burns called one of *The Twa Dogs* Luath after Cuchullin's hunting dog in *Ossian's Fingal*. Luath Press grew up in the heart of Burns country, and now resides a few steps up the road from Burns' first lodgings in Edinburgh's Royal Mile.

Luath offers you distinctive writing with a hint of unexpected pleasures.

Most UK bookshops either carry our books in stock or can order them for you. To order direct from us, please send a £sterling cheque, postal order, international money order or your credit card details (number, address of cardholder and expiry date) to us at the address below. Please add post and packing as follows: UK – £1.00 per delivery address; overseas surface mail – £2.50 per delivery address; overseas airmail – £3.50 for the first book to each delivery address, plus £1.00 for each additional book by airmail to the same address. If your order is a gift, we will happily enclose your card or message at no extra charge.

Luath Press Limited
543/2 Castlehill
The Royal Mile
Edinburgh EH1 2ND
Telephone: 0131 225 4326 (24 hours)
Fax: 0131 225 4324
email: gavin.macdougall@luath.co.uk
Website: www.luath.co.uk